Neues verkehrswissenschaftliches Journal

Ausgabe 20

DFG Final Report

The Influence of Dispatching on the Relationship Between Capacity and Operation Quality of Railway Systems

(Der Einfluss der Disposition auf den Zusammenhang zwischen Belastung und Betriebsqualität von Eisenbahnsystemen)

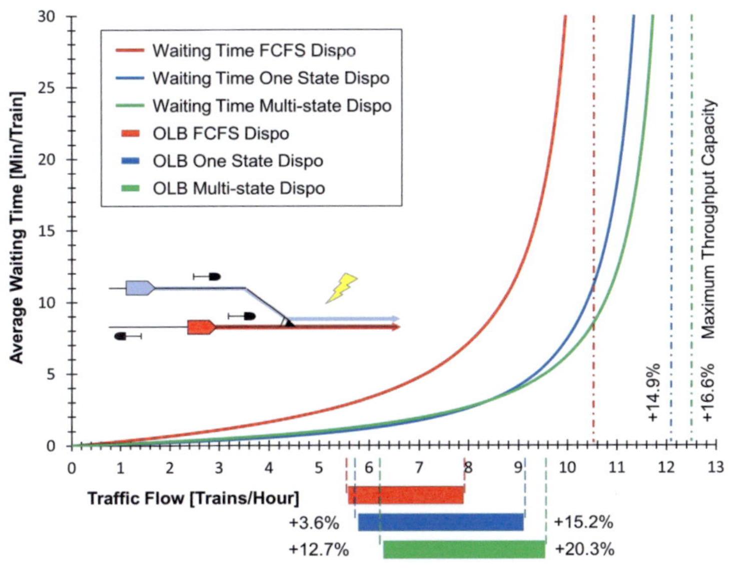

DFG Research Project MA 2326/15-1

Prof. Dr.-Ing. Ullrich Martin

M.Sc. Jiajian Liang

Institut für Eisenbahn- und Verkehrswesen der Universität Stuttgart

May 2017

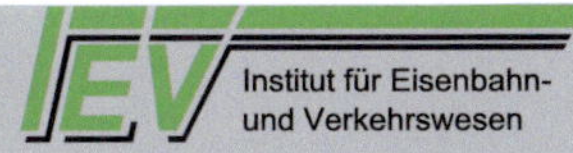

Preface

Nowadays train dispatching in railway operations is carried out in real-time and more or less manually. Therefore, the quality of the dispatching solutions highly depends on the experience of the dispatcher. Moreover, different dispatchers might have different understandings of the objectives for a given situation; making it difficult to ensure the objectives' fulfillment in real-time operations. Notwithstanding, no matter if one talks about dispatchers or dispatching assistant tools, it is clear that well defined dispatching objectives, which today are more or less general verbal expressions without quantitative indicators, would help improve the quality of the conceived solutions. In order to support dispatchers, some dispatching assistant tools have been or are being developed in which the fulfilment of dispatching objectives is based on the rescheduling of based on certain priority rules in case of deviations.

In this project, funded by the DFG (MA 2326/15-1), a new multi-scale model characterized by continuous scaling between micro-, meso- and macro-scale level has been developed. The model was developed to overcome the low accuracy of macro-scale models and the high computation complexity of micro-scale models for large areas. The model displays, through decomposition patterns, a subdivision of the network into local area to generate local solutions separately. The solutions are then coordinated on a higher level to obtain a globally feasible solution.

Through this model, simulations are concurrently carried out at the microscopic, mesoscopic, and macroscopic levels for different system states, which are considered and classified according to the number of trains per hour, the average knock-on delay and the proportion of delayed trains over the studied network within a certain time interval. Furthermore, with this model, the influences of dispatching on the relationship between capacity and operation quality, which is represented by the results of capacity research, can be evaluated systematically so that real-time operation-oriented dispatching and long-term planning-oriented capacity research can be connected.

The results of this research can be seen as an interesting contribution to the realization of automatic dispatching systems as well as a basis for further research activities.

Stuttgart, April 2017
Ullrich Martin

Table of Content

List of Figures

List of Tables

1 Introduction

Capacity research is applied to describe the relationship between capacity and operation quality within a defined investigated area of a certain infrastructure network. When simulation methods are used in capacity research, the integrated dispatching modules will influence the final results of capacity research. In the simulation of disturbed timetables with stochastic deviations, dispatching modules will be called on to solve or avoid conflicts when conflicts occur or when potential conflicts are detected. The relationship between capacity and operation quality, which is represented by the results of capacity research, includes the maximum capacity, the waiting time function [Hertel, 1992; Pachl, 2002] and the recommended area of traffic flow. These all will be different depending on the employment of different dispatching algorithms in reality and simulation of railway operation respectively. In the worst case scenario with a large number of trains, an unsuitable dispatching algorithm perhaps cannot realize the timetabled capacity, which underlines the importance of choosing suitable dispatching algorithms for the target capacity.

In order to evaluate these influences, dispatching algorithms should be clearly defined in advance. In this project, which was funded by the Deutsche Forschungsgemeinschaft (DFG Research Project), a new state-dependent dispatching algorithm, which can be adjusted according to the system states, was designed, and a classification method of system states was developed. In the further dispatching process, the system states should be identified primarily, and the most suitable dispatching algorithm can be automatically carried out. Capacity research was carried out with simulation methods. A multi-scale simulation model characterized by continuous scaling was built to calculate the waiting times based on timetables with stochastic deviations. In this model, the significant areas are simulated on a microscopic level, and the other areas are simulated on more efficient mesoscopic and macroscopic levels.

With the simulation method, three rounds of capacity analysis were carried out; in each round one of the three different dispatching algorithms were implemented: First Come First Serve (FCFS), state-dependent dispatching algorithm and a state-independent dispatching algorithm, which is designed without the consideration of system states. By comparing the results of the three rounds of capacity analysis, the influences of dispatching on the relationship between capacity and operation quality of railway systems were quantified.

Through the evaluation of the structurized influences of dispatching on the relationship between capacity and operation quality based on the theory of capacity research of railway systems, the real-time operation-oriented dispatching and long-term planning-oriented capacity research are connected. The advantages are: on one hand, suitable dispatching algorithms can increase the capacity while meeting certain operation quality requirements; on the other hand, the accuracy of capacity research can be improved. Both of them are required to increase the infrastructure exploitation rate and avoid redundant infrastructure investments.

The structure of this report is organized as follows: the related literatures will be summarized in Chapter 2; the developed multi-scale simulation model will be described in Chapter 3 (corresponding to work package 1 of the DFG project); an assessment method was created for the multi-scale simulation model in Chapter 4 (corresponding to work package 2), which could determine the proper description levels of different subareas in an investigated area; the state-dependent dispatching algorithm developed in this research will be elaborated in Chapter 5 (corresponding to work package 4); the details of the system state classification will be depicted in Chapter 6 (corresponding to work package 3); based on the state-dependent dispatching algorithm and system state classification, the influence of dispatching on the relationship between capacity and operation quality will be systematically evaluated in Chapter 7 (corresponding to work package 5); eventually, the conclusion of this research will be presented in Chapter 8.

2 Railway Dispatching and Capacity Research

In order to evaluate the influences of dispatching on the results of capacity research, railway operation processes should be simulated with a large quantity of system conditions with stochastic deviations. To generate a large amount of timetables with stochastic deviations that contains different system conditions, the software PULEIV which has been developed by IEV (Institut für Eisenbahn und Verkehrswesen der Universität Stuttgart) [Martin and Schmidt, 2010] can be used. Within the PULEIV generated timetables, initial delays of trains at the boundary of the investigation areas and original delays of trains at their origin stations located within the investigation areas are modelled by negative exponential distribution. During the simulation process of the PULEIV generated timetables, once potential conflicts are predicted, the integrated dispatching module in simulation software will be triggered to solve these conflicts. The executed dispatching actions have direct influence on simulation results, such as unscheduled waiting times of trains, which are the essential data for capacity analysis.

In different system conditions, the dispatching algorithms may have to be modified correspondingly to achieve a better performance (e.g. to fulfill a pre-defined objective function). In [Horng, 2006] it is shown that the selection of the best dispatching rule is affected by the system utilization rate in flow shop, job shop, and open shop problems. In [Vepsalainen, 1984], state dependent priority rules for scheduling, in flow shop and job shop problems are comprehensively investigated, and the scope and detail of the system state information is classified into three categories: local, indirect global, and direct global information. In [Chibana and Pierreval, 2010], system state is characterized through a series of system parameters, and a neutral network is used to describe the mapping relationship between dispatching priority rules and system states in flexible manufacturing systems.

Similarly, system states should also be considered in the rescheduling processes of railway operations, which have been rarely studied in existing research. One simple example is shown in [Martin, 1995]: trains have the same priority in case of congestion, and have different priorities in case of normal dispatching conditions. Another example is shown in [Oetting, 2010], on single-track sections if the occupation rates exceed a threshold value, trains with the same direction will be bundled in the case of deviations. In this project, the system states will also be classified according to the number of trains per hour, the average knock-on delay and the proportion of delayed trains over the studied network within a certain time interval.

In the design of dispatching algorithms, generally three types of dispatching approaches could be considered: simulative, analytical and heuristic approach (similar classification also can be found in [Corman and Meng, 2013], [Martin, 2002] and [Cui, 2010]. With the simulative approach the operation process is simulated as in reality. During the simulation process, traffic situations are predicted periodically. Once a conflict is detected, the integrated dispatching module will be called on to solve the conflict, and the processing technique of the simulation model (i.e. synchronous and asynchronous) plays an important role in the mechanism of conflict resolution. For synchronous simulations, train movements are updated progressively and interacting with each other immediately [Siefer, 2008]. Therefore, the integrated dispatching module imitates dispatchers to solve conflicts chronologically. For instance, a series of dispatching measures, which include overtaking, replatforming, dwell time extension and so on, are implemented in the software RailSys [RMCon, 2007]. When potential conflicts are detected during the simulation process, suitable dispatching measures will be performed. For asynchronous simulations, train paths are inserted in the time-distance diagram in sequence of priority. A dispatching assistant tool – ASDIS (Asynchronous Dispatching) – was developed in the research project DisKon [Shaer et al., 2005]. After a train group (with the same priority) was inserted, the conflicts among the equal or higher-ranking trains should be resolved chronologically.

With analytical approaches, train operations are modelled as mathematical equations, such as mixed integer linear programming [Corman et al., 2016], queuing theory (e.g. [Marinov and Viegas, 2011] and alternative graph model (e.g. [Corman et al., 2011] and [D'Ariano and Pranzo, 2008] and so on. For overviews of mathematical models it is referred to [Alwadood et al, 2012], [Cacchiani et al, 2013] and [Corman and Meng, 2013]. It is very complex and time-consuming to solve the analytical models with exact method. The heuristic approach is a good alternative, which can balance the solution quality and computation time.

Heuristic approaches include a wide range of algorithms, such as tabu search, simulated annealing, swarm intelligence and so on. In [Cui, 2010] a macroscopic railway dispatching optimization model was developed based on tabu search, in which changes of train sequences on open track sections are defined as basic move operations. In [D'Ariano, 2008] tabu search is used to optimize train paths, and the Branch and Bound algorithm to optimize the train sequences on infrastructure resources. In [Samà et al., 2016] the ant colony (an example of swarm intelligence) is chosen as the basis of the dispatching optimization algorithm, and it is aimed to solve the train routing selection problem. Many heuristic or metaheuristic algorithms are available to be used as the basis of the dispatching algorithm in railway operation. A widely used heuristic algorithm – greedy algorithm – was preferred in this research. The local search mechanism of the greedy algorithm is the basis of many heuristic and metaheuristic algorithms. Therefore, the specific local search mechanism developed in this section can be easily implemented under the framework of other heuristic or metaheuristic algorithms.

In the design phase of simulation models, special attention should be paid to the performance issues. In general, simulation models can be classified into three types – microscopic, mesoscopic and macroscopic models – according to their levels of detail of the description. Microscopic simulation software have been widely used in both industry and academia, such as the software RailSys developed by Leibniz Universität Hannover and Rail Management Consultant [RMCon, 2016], OpenTrack developed by Eidgenössische Technische Hochschule Zürich [OpenTrack, 2016] and LUKS developed by VIA Consulting and Development GmbH [VIA-Con, 2016]. Microscopic simulation models simulate railway operation processes in investigated areas very precisely, and the accuracy of simulation results is the highest. However, due to its high computational complexity, the computation effort within a large investigated area is not yet acceptable. Therefore, mesoscopic and macroscopic simulation models have to be adopted to reduce complexity in such cases.

According to the current state of art, there is no universal accepted simplification method on the mesoscopic level. The existing mesoscopic models are customized based on different application contexts. To evaluate the impact of freight train operations in a railway network, a mesoscopic simulation model based on queening system is developed in [Marinov and Viegas, 2011]. In this model, an infrastructure network is decomposed into interconnected components, such as stations, yards and open tracks between stations. For instance, a passenger station is modeled as one storage areas (i.e. queue) and one work centers (i.e. server) in each running direction. The capacity of the storage area is restricted by the number of the station platform tracks. The service time in the work center refers to the running time of a train through the considered station. In [Fabris et al., 2014] a different mesoscopic model is developed to generate railway timetables. The interlocking areas in stations are modeled with matrices of station routes and their compatibility. The other parts of the network are modeled in the same manner as microscopic models. Furthermore, running times are also estimated with train motion equations. This mesoscopic model can be treated as a quasi-microscopic model. In [Radtke, 2014] a down-scaling approach is introduced, and the artificial infrastructure network generated based on a pre-given macroscopic infrastructure network is defined as a mesoscopic infrastructure network. From the literatures mentioned above, it can be seen that the levels of detail of the mesoscopic infrastructure models differ significantly. A station can be abstracted into a point (e.g. work center and storage area in [Marinov and Viegas, 2011]), or depicted almost as accurately as the microscopic level (e.g. only interlocking areas are abstracted in [Fabris et al., 2014]). In order to achieve better applicability, a mesoscopic infrastructure model characterized by continuous scaling was developed in this project.

Macroscopic models are capable of simulating a large investigated area very efficiently. A macroscopic infrastructure network can be described with the node-link-model elaborated in [Radtke, 2014]. Nodes represent stations and junctions, and links represent open track sections between nodes. Due to the high abstraction level of macroscopic models, additional operation constraints should be defined in order to regulate train movements as in reality. For instance, running times of each train group on all relevant links and the minimum line headway for each pair of train groups that run on the same link should be prepared in advance, and train runs on open track sections are separated with minimum line headways (e.g. [Kettner et al., 2003] and [Cui, 2010]). Besides minimum line headways, the other constraints on open track sections are classified into three levels in [Kecman et al., 2012]: both conflicts between train runs that on the same open track section and that from different open tracks on merging or intersecting points are ignored; only conflicts between train runs that on the same open track section are considered; both types of conflicts are considered. The evaluation results in [Kecman et al., 2012] showed that train reordering actions were quite well captured in the macroscopic model with the strictest constraints (the last case), but the average knock-on delay estimated with this model differs significantly from the results estimated with a microscopic model.

Taking the advantage of each individual simulation models (i.e. microscopic, mesoscopic and macroscopic models), a multi-scale simulation model characterized by continuous scaling was developed in this project, which was first proposed in [Cui and Martin, 2011]. In this model, simulations on the microscopic, mesoscopic and macroscopic levels are allowed to be executed concurrently. This enables the core regions in an investigated area to be accurately simulated on the microscopic level, while the surrounding regions to be efficiently simulated on the mesoscopic or macroscopic levels. With this new simulation mechanism, accuracy and computation complexity could be well balanced.

3 Scaling Method for the Developed Multi-scale Model

In order to evaluate the influence of the different dispatching algorithms, a multi-scale simulation model was developed based on continuous scaling [Cui and Martin, 2011]. There are two directions of scaling: from the microscopic level to the macroscopic level (bottom-up approach) and from the macroscopic level to the microscopic level (top-down approach). In the upscaling process, the most detailed data on the microscopic level is always available, and the most accurate simulation results can be easily obtained with the aid of the microscopic model. However, due to the high computational complexity on the microscopic level, the computation time on a large study area is not yet acceptable. Therefore, following the same structure of the microscopic model, a mesoscopic and a macroscopic model were developed in this approach in order to improve calculation efficiency. The details of the upscaling method will be elaborated in Section 3.1.

In practice, accurate data on microscopic level is not always available due to various reasons, thus it is sometimes necessary to estimate the detailed infrastructure attributes on a lower level based on the abstracted attributes on an upper level. To this end, a down scaling method was developed and will be described in Section 3.2.

3.1 Upscaling Method for the Multiscale Model

Under the framework of CUI's model developed in [Cui, 2010], infrastructure resources, simulation performers and simulation tasks (correspond to infrastructure, train and timetable in reality) are modeled as the fundamental components constructing the extended simulation model, and the interaction mechanism among the components are modeled as the workflow of this simulation model. Depending on the requirement of a description level (the microscopic or mesoscopic or macroscopic level), the attributes of the components will be depicted in different levels of detail. Meanwhile, the structure of the workflow must be kept the same. A consistent workflow structure is the prerequisite to ensure that simulations on different description levels could be executed concurrently on the same platform.

In [Cui, 2010] the workflow is designed as a series of single processing steps driven by a certain time interval as shown in Figure 3-1, and three activities are carried out successively in each step: request resources, allocate resources and proceed with simulation tasks. Along with the iterative execution of the single processing step during the simulation process, the execution time is accumulated step by step as in reality. The simulation can be terminated when the execution time reached a pre-configured time or all trains have either arrived at their destinations or run out of the investigated area.

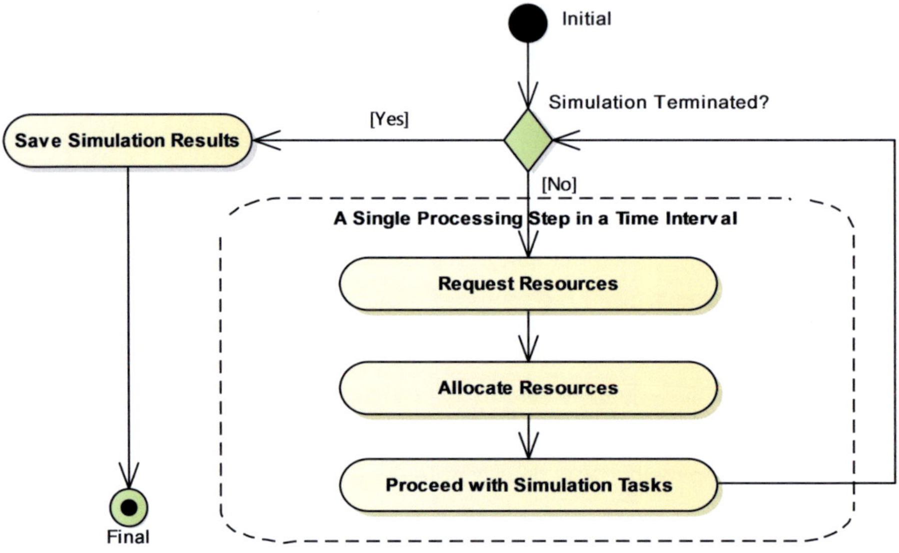

Figure 3-1: The Workflow of Synchronous Simulation [Cui, 2010]

Because the components of the simulation model – infrastructure resources, simulation performers and simulation tasks – are abstracted to different extents on different description levels, the detailed procedures involved in each activity may have to be accordingly modified for different description levels. Therefore, to carry out an activity in a single processing step, the current description level of the considered train should be determined primarily, based on which the right procedures can be correspondingly executed. The current description level of a train updates automatically in the simulation process as the train head has reached a region with a different description level. The components and workflow on the microscopic, mesoscopic and macroscopic levels will be separately elaborated in Section 3.1.1 to Section 3.1.3.

3.1.1   The Components on the Microscopic Level

3.1.1.1   Infrastructure Resources

The link and node model depicted in [Radtke, 2014] has been widely used to describe complex railway infrastructure network. Wherever an attribute of railway lines varies, a node has to be inserted to indicate the attribute variation (e.g. signals, points, timing points etc.). A link connects two adjacent nodes, and stores all relevant information (e.g. speed, gradient, radius etc.). Given an infrastructure network depicted with the link and node model, the software PULEIV developed by IEV [Martin et al. 2008a; Martin et al. 2008d; Martin et al. 2008c] can automatically decompose the infrastructure network into individual basic structures[1]. A basic structure is the maximum occupation unit allowed to be occupied only by one train simultaneously on the microscopic level. Basic structures are used as a basic unit to be requested, allocated and released in the microscopic model. Furthermore, basic structures can be classified into two categories: junction-type basic structures (containing turnouts or crossings), and non-junction-type basic structures (containing only tracks). The process of basic structure partitioning based on an exemplary infrastructure is illustrated in Figure 3-2. As shown in the figure, once a basic structure is determined, new edges will be created to replace the included links. The attributes of the new edge can be derived from the included links.

---

[1] In [Martin and Li, 2014] a basic structure is defined as follows: "a basic structure is an interrelated part of an infrastructure network. As a non-directional occupancy element, it is bounded by main signals, signal or route releasing points or the boundary of the investigated area."

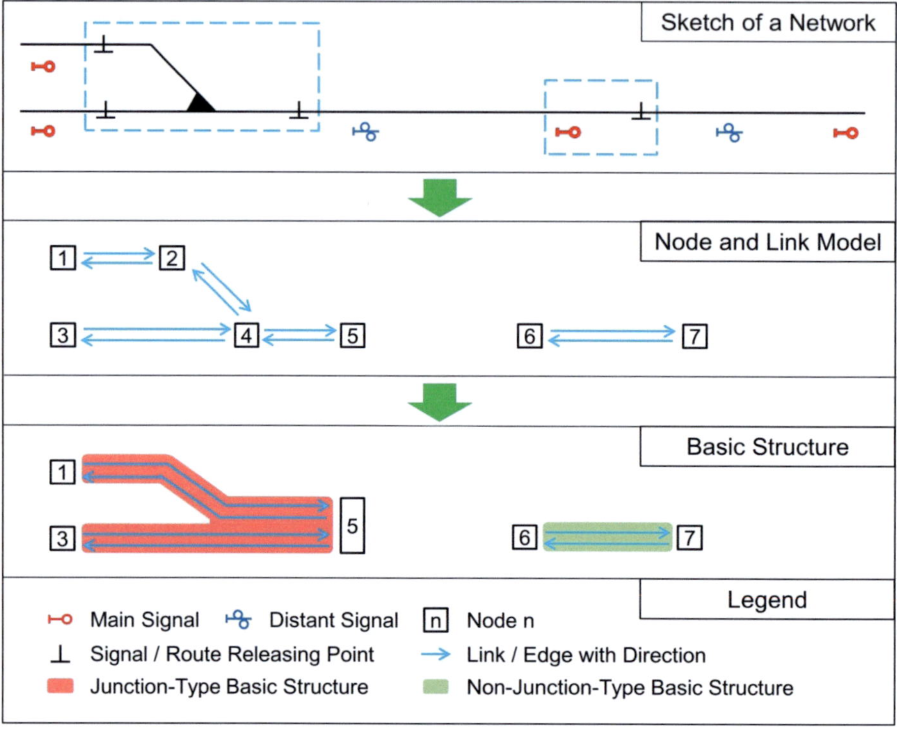

Figure 3-2: An Example of Basic Structure Partitioning Algorithm

As the input data for an infrastructure network, the attributes of each edge (such as length, permissible speed, gradient and radius) and the attributes of each basic structure (such as included edges, and whether it is a free resource[2]) should be entered before a simulation is executed.

The input data on block sections and train paths should also be prepared after the determination of the edges and basic structures of the infrastructure network. A block section is composed of an array of edges in the same direction, and the corresponding basic structures are stored as an attribute of the block section (Figure 3-4). Furthermore, the automatic train protection system (ATP) of the block section should

---

[2] A free resource is defined as a virtual resource extended from the investigated infrastructure network for the convenience of modeling, and any train is free to enter such resources at any time (no need of movement authority).

also be configured. Two ATP systems – intermittent ATP and continuous ATP are implemented in this approach. In the intermittent ATP system, two data transmission points - the main signal and distant signal - are taken into account. The main signal indicates the occupancy information about the block section behind the signal, and the distant signal indicates the approach information for the next signal (Figure 3-3). Moreover, block overlaps are also considered for the intermittent ATP system. A block overlap is represented as a series of basic structures behind the corresponding block section. In the continuous ATP system, the cab signaling system is used to transmit data continuously, and block markers are only used to indicate the physical boundaries of block sections. For more details of the intermittent and continuous ATP systems, it is referred to [Pachl, 2002].

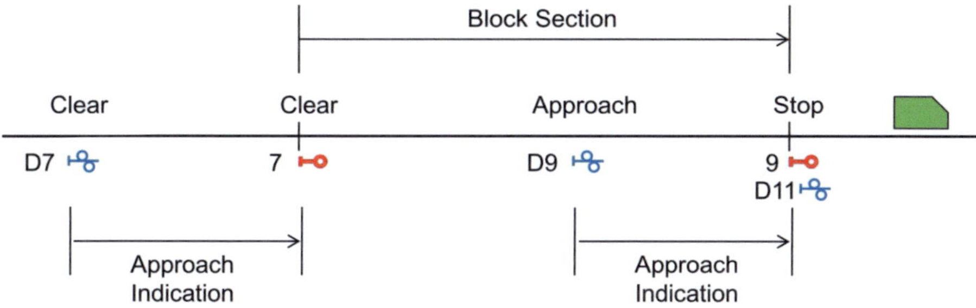

Figure 3-3: Intermittent ATP System

A train path is composed of a sequence of block sections, and always starts and ends up with block sections composed of exclusively free resources, where train movements start and terminate. The block sections and train paths defined on an exemplary infrastructure network are illustrated in Figure 3-4.

Besides the attributes of edges and basic structures, the attributes of each block section (such as the included edges and corresponding basic structures, ATP system and distant signal distance) and the attributes of each train path (such as included block sections) also should be entered as the input data for the infrastructure network.

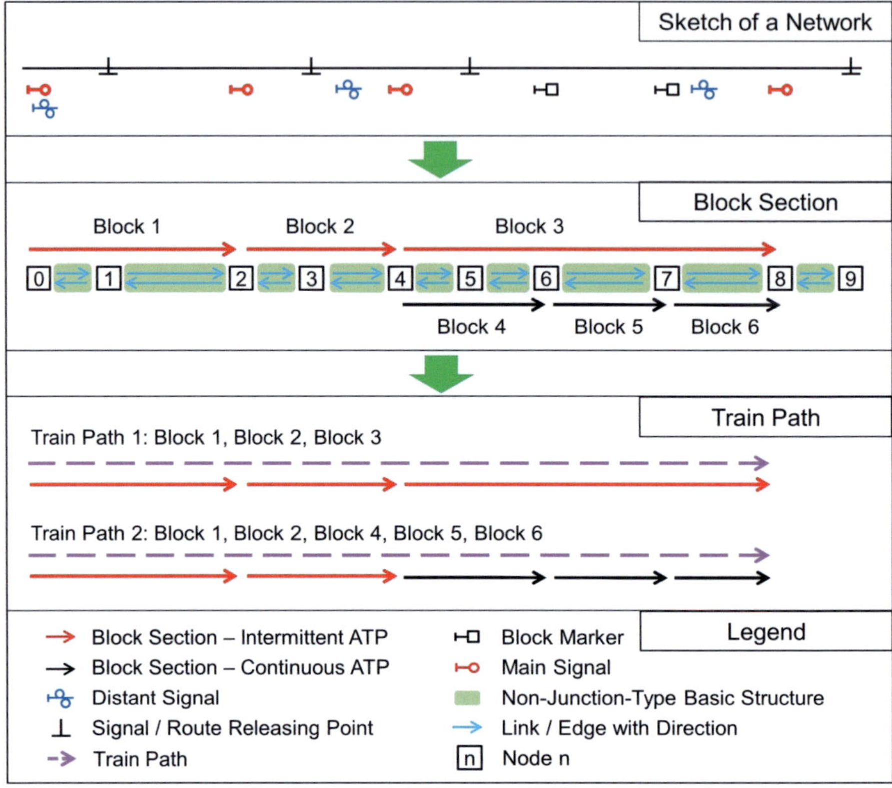

Figure 3-4: An Example of Block Section and Train Path

3.1.1.2 Simulation Performers and Simulation Tasks

In the simulation model, only train movements rather than shunting movements are considered, and hence simulation performers are configured as trains. A train is defined as a locomotive or self-propelled vehicle, along or coupled to one or more vehicles operating on main tracks [Pachl, 2002]. To describe train dynamics, the detailed physical and mechanical attributes of trains should be entered, such as train type, maximum speed of the train, mass of the traction unit and vehicles and so on. Based on the attributes of a considered train and the relevant infrastructures, the movement behavior of the train can be accurately depicted. Three behavior sections - acceleration section, constant movement and braking section - are implemented in the microscopic model. Theoretically, the movement behavior of a train in a time interval can be any combination of these three sections. The coasting section can also be implemented. Therefore, to estimate the forward distance of a train in a certain time interval, the involved behavior sections should be determined primarily according to the attributes of the relevant infrastructures and the considered train and the current operational constraints (e.g. scheduled or unscheduled stops). The forward distance of the train in each involved behavior section can be calculated by solving train dynamics equations (for details of train dynamics equations it is referred to [Brünger and Dahlhaus, 2014]). The sum of these results is the forward distance of the train in the considered time interval. With the assistance of forward distance estimation, the position of the train is updated step by step, and infrastructure resources are accordingly requested, occupied and released following the simulation workflow. The workflow on the microscopic level will be elaborated in Section 3.1.2.

A simulation task describes the scheduled operations of a train in the investigated area. Based on a pre-given timetable, simulation tasks can be easily derived. A simulation task should contain the following information: the available train paths for the considered train, the scheduled stops along each train path, the scheduled departure and dwell time at each stop and specific stops for turnaround tasks. Three are three states in the entire lifecycle of a simulation task: created, running and terminated. At the beginning of a simulation, the states of all simulation tasks are initialized as "created". During the simulation process, as soon as the scheduled departure time of a train at the initial stop or the boundary of the investigated area is reached, the state of the corresponding simulation task is set as "running". When a train has arrived at its destination (the block section composed exclusively of free resources at the end of its train path), the state of the corresponding simulation task is set as "terminated". Thus it is only necessary to perform the simulation tasks with the state of "running" in the simulation process.

3.1.2   The Workflow on the Microscopic Level

3.1.2.1   Request Resources

In each single processing step, the requested resources of every train should be determined. Under the regulation of different ATP systems, the trigger mechanisms of resource requirements vary as well. So the intermittent and continuous ATP systems will be separately discussed in this section. To determine the resource requirement in each ATP system, the status of trains, such as current speeds and positions, should be considered. For instance, in the intermittent ATP system, only when the head of a train is going to pass a distant signal or main signal, the train has chance to request new resources in the current time interval.

In addition, resource requirement is also restricted by actions of trains. On condition that a train is approaching a scheduled stop (the block section designated as the scheduled stop has been allocated to the train), even though the head of the train is going to pass a distant signal, new resource requirement should not be allowed. So actions of trains are introduced to support the determination of train status. Four actions are defined herein, including "Run", "Pre-stop", "Stop" and "Turnaround". As the attributes of trains are loaded at the beginning of a simulation, the actions of all trains are initialized to "Run". In the simulation process, as soon as a train obtained a scheduled stop, the action of the train will be updated to "Pre-stop". Afterward, the train will be brought to a halt on the scheduled stop, and thereupon the action of the train will be updated to "Stop". When scheduled departure time is reached (schedule dwell time must be fulfilled), it is necessary to check whether a turnaround task is arranged according to the simulation task. If so, the turnaround task should be executed at first, and then the action should be updated to "Turnaround"[3]; otherwise, the action should be directly updated to "Run". Obviously, only in case of "Run" and "Turnaround" the corresponding train may have to request new resources.

With consideration of the ATP systems and status of trains, 8 request-cases are summarized to cover all situations in which resource requirement should be triggered. Four of them belong to the intermittent ATP system, and the others belong to the continuous ATP system. Excepting these 8 request-cases, resource requirement is unnecessary. The process of resource requirement is illustrated in Figure 3-5. The boundary conditions of each request-case are clearly defined according to the employed ATP system and the status of the considered train. For each ATP system, one representative request-case (Request-Case 1 and Request-Case 5) will be elaborated in the following context (for the details of the other request-cases it is referred to [Liang, 2017]).

---

[3] The action "Turnaround" means that the scheduled turnaround task has been completed, and the train is allowed to request new resources beyond the scheduled stop.

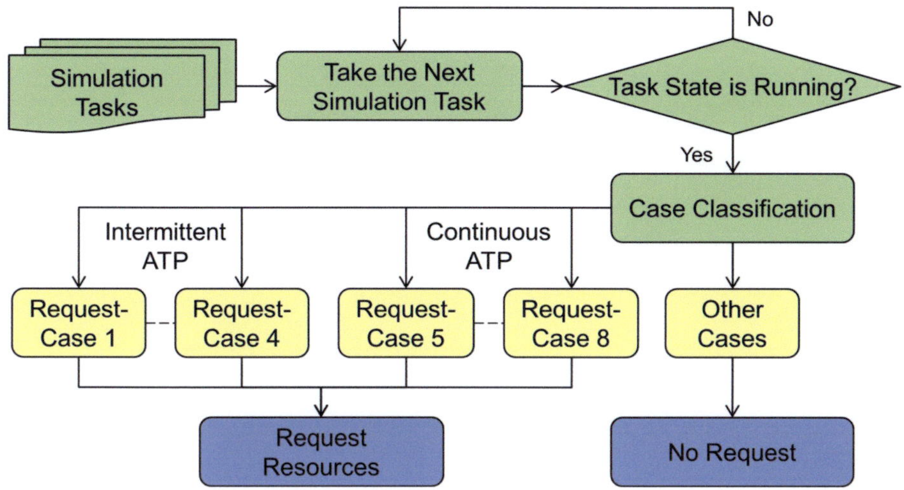

Figure 3-5: Process of Resource Requirement

The boundary conditions of **Request-Case 1** are defined as follows:

- the intermittent ATP system is employed AND[4]
- the action of the considered train is "Run" AND
- the current speed of the considered train is not equal to zero AND
- the current position of the train head locates between the entrance signal of the last block section and the distant signal for the next block section (Figure 3-6) AND
- at the end of the current time interval, the train head will pass the distant signal for the next block section (Figure 3-6).

If the boundary conditions of Request-Case 1 are fulfilled in a time interval, resources requirement will be triggered for the corresponding train. As shown in Figure 3-6, the basic structures included in the next block section will be requested. Furthermore, if the next block section belongs to the intermittent ATP system, the overlap of the next block section should also be included in the requested resources.

---

[4] "AND" refers to the logical AND operator.

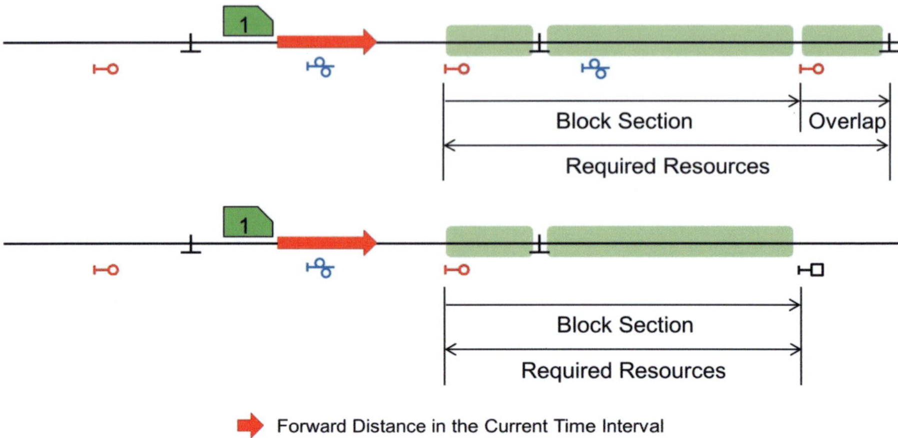

Figure 3-6: Resource Requirement – Request-Case 1

The boundary conditions of **Request-Case 5** are defined as follows:

- the continuous ATP system is employed AND

- the action of the train is "Run" AND

- the current speed of the train is not equal to zero AND

- the nominal stopping point is going to pass a block marker in the current time interval AND

- the train head will not enter the next block section at the end of the current time interval OR[5] the next block section still belongs to the continuous ATP territory[6].

The nominal stopping point is used to detect resource requirements for the continuous ATP system. Given the relevant attributes of a considered train and the relevant infrastructures at a certain time instance, the braking curve of the train can be deduced with train dynamic equations (e.g. red curve in Figure 3-7). The nominal stopping point refers to the endpoint of the braking curve. All the block sections located

---

[5] "OR" refers to the logical OR operator.

[6] Request-Case 5 only deals with resource requirement in the continuous ATP territory. If the train head will enter a block section in an intermittent ATP territory, the resource requirement mechanism of the intermittent ATP system should also be taken into account.

between the position of the train head and the nominal stopping point should be blocked for the train, which is the basic requirement of safe train separation.

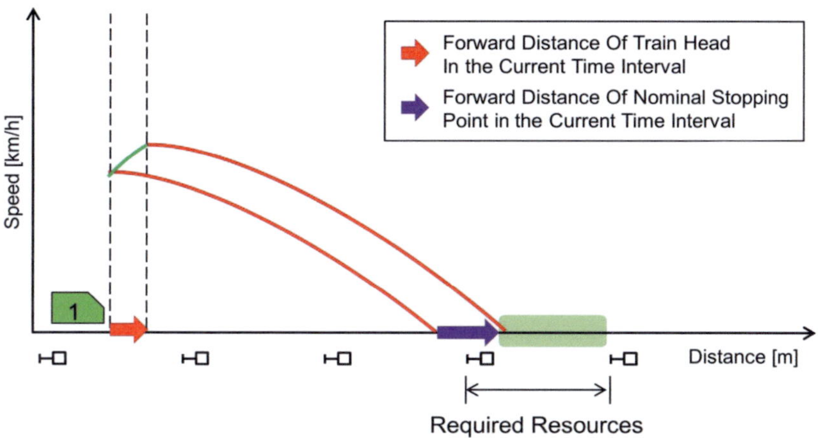

Figure 3-7: Resource Requirement – Request-Case 5

If all boundary conditions of Request-Case 5 are fulfilled, the basic structures included in the block section behind the block marker will be requested by the train.

3.1.2.2 Allocate Resources and Proceed with Simulation Tasks

After all resource requirements have been determined, all resource-requester pairs should be checked in turn by the conflict-free and deadlock-free test. For the conflict-free test, if any of the basic structures requested by a train is occupied by another train[7], the resource requirement of the train can be directly denoted as conflict. Resource requirements with conflicts will be directly rejected, and do not have to be further checked with the deadlock-free test. For deadlock-free test, the deadlock avoidance algorithm based on the Banker's algorithm developed in [Cui, 2010] is employed herein (for details of the algorithm it is referred to Chapter 4 in [Cui, 2010]). Only the resource requirement successfully passed conflict-free and deadlock-free test will be allocated to the corresponding train. At the end of the resource allocation

---

[7] The free resource is an exception. By definition any train at any time can be authorized to enter free resources.

procedure in a time interval, all resource-requester pairs have to be cleared, since the resource requirement will very likely be different in the next time interval.

Based on the results of resource allocation, the attributes of infrastructure resources, trains and simulation tasks have to be properly updated in different conditions. According to the characteristics of different conditions, 14 proceed-cases are summarized for the process of proceeding with simulation tasks: 8 of them are defined for the intermittent ATP system, and the others are defined for the continuous ATP system[8] (Figure 3-8). One representative proceed-case (Proceed-Case 1 and Proceed-Case 9) will be elaborated for each ATP system in this section (for the details of the other proceed-cases it is referred to [Liang, 2017]).

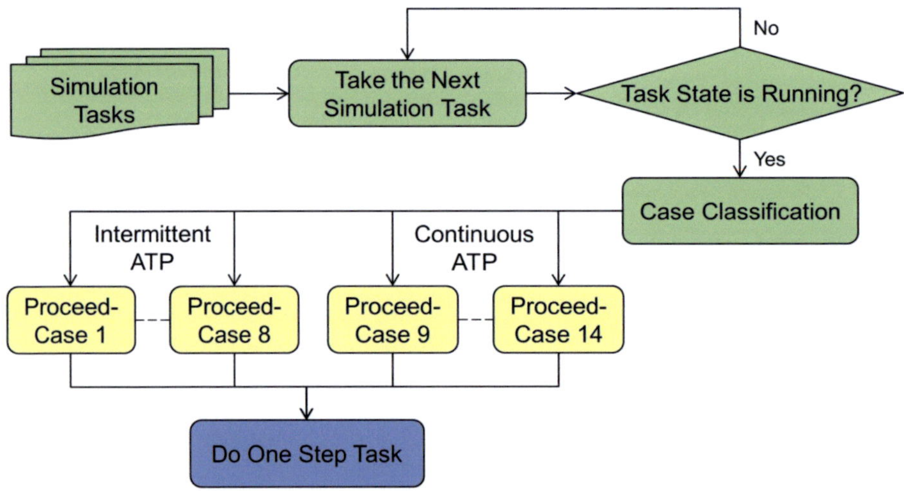

Figure 3-8: Process of Proceeding with Simulation Tasks

The boundary conditions of **Proceed-Case 1** are defined as follows:

—   the intermittent ATP system is employed AND

—   the action of the train is "Run" OR "Pre-stop" AND

—   the current speed of the train is not equal to zero AND

---

[8] The cases classified in this section are used to support the proceeding with simulation tasks. Therefore, they do not necessarily have corresponding relationship with the cases classified for resource requirement.

– the current position of the train head locates between the entrance signal of the last block section and the distant signal for the next block section (Figure 3-9) AND

– at the end of the current time interval, the train head will pass the distant signal for the next block section (Figure 3-9).

Proceed-Case 1 is further divided in two subcases. In the first subcase, the action of the train is "Run" and requested resources are obtained. The following procedures should be carried out for the first subcase:

– If the newly obtained block section is a scheduled stop, the action of the train should be updated to "Pre-stop".

– If the permissible speed of the next block section is reduced, a break application point[9] (P2 in Figure 3-9) for speed reduction should be determined.

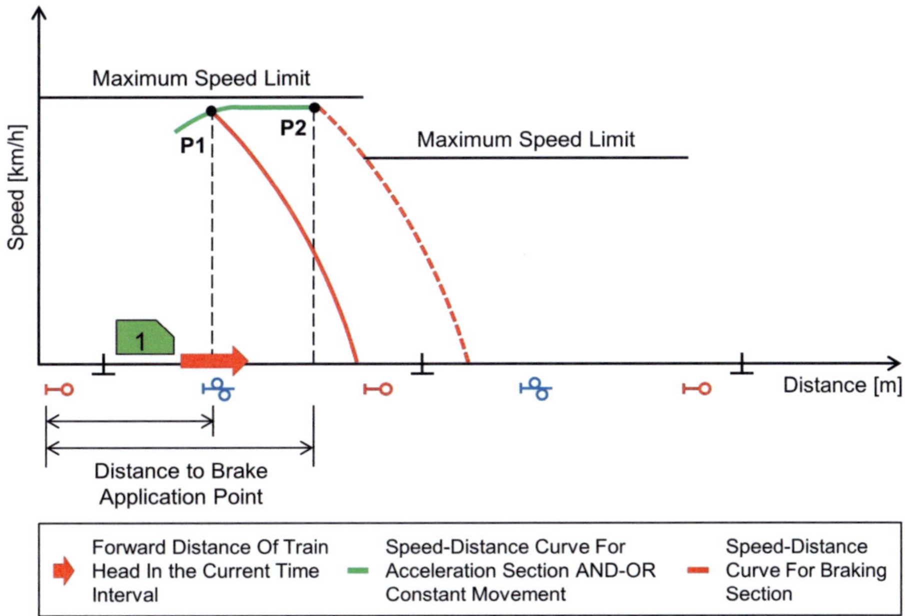

Figure 3-9: Proceed with Simulation Tasks – Proceed-Case 1

---

[9] The function of a brake application point is to indicate the right position for a train in an intermittent ATP territory, from which the train must brake until it stopped or entered the next block section with a safe speed. After the mission of a brake application point is accomplished, it must be removed.

In the section subcase, the action of the train is "Pre-stop" or the requested re-sources were rejected. A brake application point for stop (P1 in Figure 3-9) should be calculated and saved. Once a new brake application point is inserted, the forward distance in one time interval should be reestimated. Depending on the estimated for-ward distance, the following procedures should be executed in the current time inter-val for both subcases:

- update the current position, current speed and current physically occupied block sections of the train;
- release the basic structures behind the train rear.

Last but not least, if the newly obtained block section belongs to a continuous ATP territory, the signaling system of the train should be switched. In the next time interval, train movement will be regulated by the continuous ATP system. So, if a the brake application point for speed reduction has been inserted in the current time interval, it should be replace with a stopping point for permissible speed reduction (e.g. P2 should be replace with P3 as shown in Figure 3-10). Similar to the brake application point for the intermittent ATP system, a stopping point for permissible speed reduc-tion is used to indicate when a train should start to brake in the continuous ATP sys-tem.

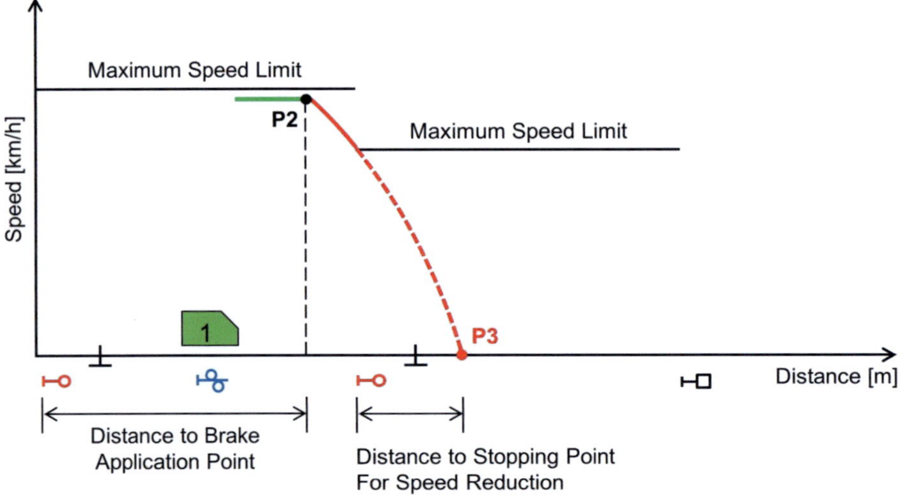

Figure 3-10: Switching of ATP System – From Intermittent ATP to Continuous ATP

The stopping point[10] is a general concept to control train speed in case of permissible speed reduction or scheduled or unscheduled stops in the continuous ATP system. It contains two subclasses: stopping point for permissible speed reduction and temporal stopping point. A stopping point has three attributes: the source block section, the target block section and the distance between the entrance block marker of the target block section and the stopping point. For the stopping point for permissible speed reduction, the source block section represents where permissible speed reduction firstly occurred, and the target block section represents where the stopping point locates (as shown in Figure 3-11). If the nominal stopping point of a train is going to pass the stopping point for permissible speed reduction, the train has to start braking in the current time interval. Once the train entered the source block section, the stopping point for speed reduction ought to be removed. The details of temporal stopping point will be explained later in this section.

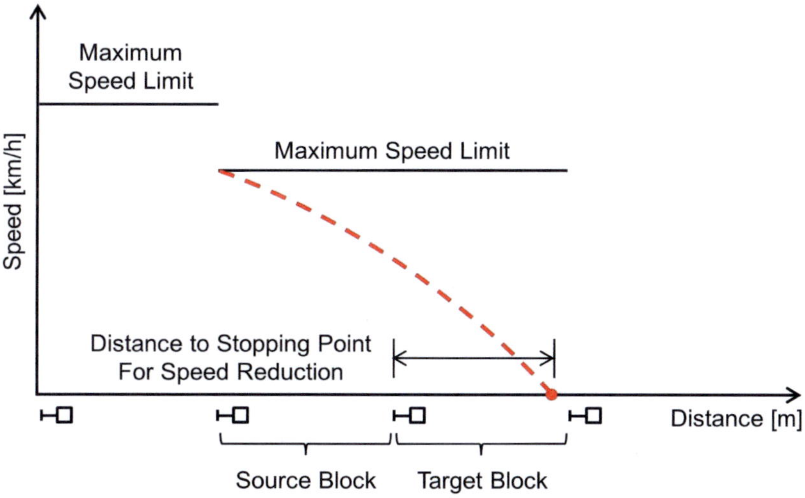

Figure 3-11: Stopping Point for Maximum Speed Limit Reduction

---

[10] The stopping point is different from the aforementioned nominal stopping point of a train. The nominal stopping point is used to detect new resource requirement, while the stopping point is used to support train speed control in the continuous ATP system.

The boundary conditions of **Proceed-Case 9** are defined as follows:

- the continuous ATP system is employed AND
- the action of the train is "Run" OR "Pre-stop" AND
- the current speed of the train is not equal to zero AND
- the train head will not enter the next block section at the end of the current time interval OR the next block section still belongs to the continuous ATP territory.

Proceed-Case 9 also contains two subcases. In the first subcase, the nominal stopping point of a train is going to enter a new block section. On condition that new resources were requested and allocated to the train (the action of the train must be "Run"), the following procedures should be carried out:

- If the permissible speed of the newly obtained block section is reduced, a stopping point for speed reduction should be determined;
- If the newly obtained block section is a scheduled stop, the action of the train should be updated to "Pre-stop";

On condition that the requested resources were not allocated to the train (the action of the train must be "Run"), the train has to decelerate in the current time interval. To this end, a temporal stopping point is introduced. As the name suggests, the life cycle of a temporary stopping point is set to one time interval, since the results of resource allocation may be different in the next time interval. The source block of a temporal stopping point can be set as null, and the target block is the last operationally occupied block section of the train. The temporal stopping point is just located at the end of the target block section (Figure 3-12). With the help of the temporary stopping point, even in the worst case the train can be brought into an unscheduled stop in the target block section. In principle, the case of "Pre-stop" (resource requirement is not allowed) can be treated in the same manner. By means of a series of temporal stopping points in the following time intervals, eventually the train will be brought to a halt at the scheduled stop.

In the second subcase, the nominal stopping point of the train is not going to enter a new block section, so no resource was requested. Only regular procedures should be executed, which is also necessary for the first subcase:

- update the current position, current speed and current physically occupied block sections of the train;
- release the basic structures behind the train rear.

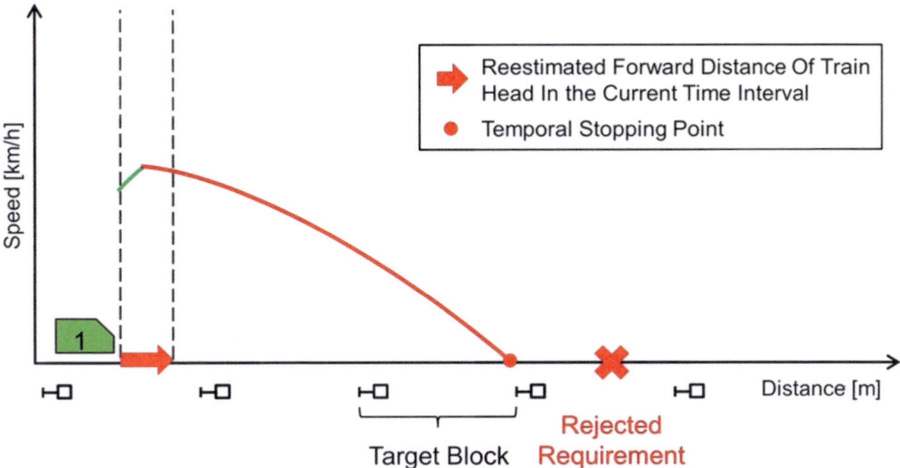

Figure 3-12: Temporal Stopping Point in Continuous ATP Territory

3.1.3 The Components and Workflow on the Mesoscopic Level

On the mesoscopic level, the components[11] of the simulation model - infrastructures and trains - will be simplified in order to improve the calculation efficiency of the simulation model. On the most detailed mesoscopic level, the infrastructure network is kept the same as the microscopic level, and basic structures are used as the minimum occupation unit. As the abstraction level increases, adjacent two basic structures will be combined into larger mesoscopic occupation unit to simplify infrastructure network topology (especially in interlocking areas). On further abstraction levels, two adjacent larger mesoscopic occupation units can also be aggregated. Regardless of the degree of abstraction block sections will continuous to be used. So, if any connection node between two adjacent occupation units is a main signal or block marker, the aggregation is not allowed. The aggregation of occupation units may in-

---

[11] The component – simulation task (corresponding to timetable in reality) – cannot be further simplified, because the information contained in a simulation task (e.g. scheduled stops and dwell times) are already very limited, and they are indispensable to simulate train runs as scheduled.

fluence of the accuracy of the model. In the up-scaling process the aggregations with small influences should have relative higher priority than the ones with big influences. The calculation method of aggregation accuracy will be elaborated in Section 4.2.

With regard of depiction of trains on the mesoscopic level, maximum constant average running time is used herein. Under condition that only infrastructure network topology is simplified but running time is still calculated with train motion equations, the improvement of calculation efficiency is very limited. Whereas, under condition of constant running time, not only forward distances can be very fast estimated, but the procedures involved in the workflow can be simplified to a large extend as well.

At a certain time instance, the maximum allowed speed of a train is the minimum value among the maximum speed of the train and the permissible speeds of the current block sections physically occupied by the train. The forward distance is the product of the maximum allowed speed[12] and the length of a time interval.

To ensure safe train separation on the microscopic level, the braking distance is considered in the procedure of resource requirement, for instance, distant signals in the intermittent ATP system and nominal stopping points in the continuous ATP system are used to trigger resource requirement. Even though the requested resources are not allocated to the train, the train is able to start braking in time and stop before a stop signal. However, constant running time is employed on the mesoscopic level, which implies that the train can always be forced to stop before a stop signal without restriction of speed. Therefore, distant signals and nominal stopping points become unnecessary in the procedure of resource requirement on the mesoscopic level. So, on the mesoscopic level, only if a train is physically going to enter a new block section, resource requirement is executed. Two request-cases are defined to trigger resource requirement on the mesoscopic level, and one representative request-case will be elaborated in this section (for the details of the other request-cases it is referred to [Liang, 2017]).

---

[12] Minimum running time is considered herein. If recovery times exist, the maximum allowed speed should be accordingly adjusted.

The boundary conditions of **Request-Case 1** are defined as follows:

- the action of the train is "Run" AND
- the train head will enter the next block section at the end of the current time interval.

In this case the basic structures belonging to the next block section will be requested by the train (Figure 3-13). As shown in the following figure, overlaps will no longer be considered for resource requirement, because they may disappear due to aggregation of occupation units on further abstraction levels. Moreover, without consideration of overlaps, the mechanism of resource requirement in the intermittent ATP system becomes entirely identical as that in the continuous ATP system. Therefore, the model can be further simplified. No matter in the intermittent or continuous ATP system, resource requirement can be treated in the same manner.

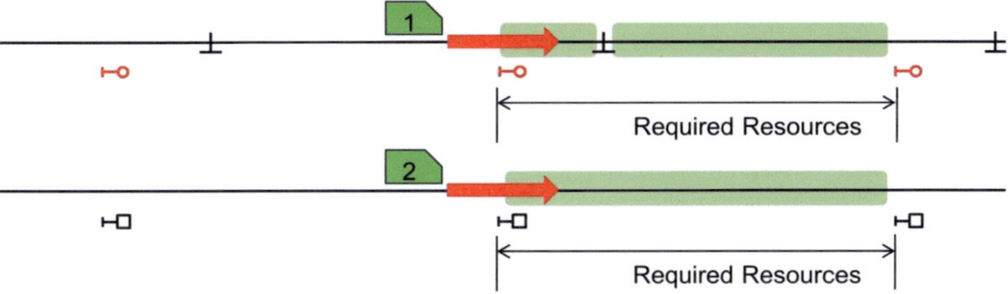

Figure 3-13: Request Resources – Request-Case 1 on the Mesoscopic Level

After the procedure of resource requirement, all resource-requester pairs should be check with conflict-free and deadlock-free test as described in Section 3.1.2.2. Resource requirements could happen on any description level (microscopic, mesoscopic and macroscopic), but resource allocation must be treated in a centralized manner.

With regard of proceeding with simulation tasks, four proceed-cases are defined, and one representative proceed-case will be explained herein (for the details of the other proceed cases it is referred to [Liang, 2017]). The boundary conditions of **Proceed-Case 1** are defined as follows:

- the action of the train is "Run" or "Pre-stop" AND
- the train will enter the next block section in the current time interval

This case contains two subcases. In the first subcase, the next block section (the action of the train must be "Run") was allocated to the train. The following procedures should be carried out:

- If the newly obtained block section is a scheduled stop, the action of the train should be updated to "Pre-stop";
- update the current position, current speed and current physically occupied block sections of the train (if the permissible speed is changed, the forward distance should be reestimated as shown in Figure 3-14);
- release the basic structures behind the train rear.

In the second subcase, the next block section was not allocated to the train (the action of the train could be "Run" or "Pre-stop"). The position of the train head should be updated to the end of the last current block section, and the basic structures behind the train rear should be released. Last but not least, if the action of the train is "Pre-stop", it should be updated to "Stop".

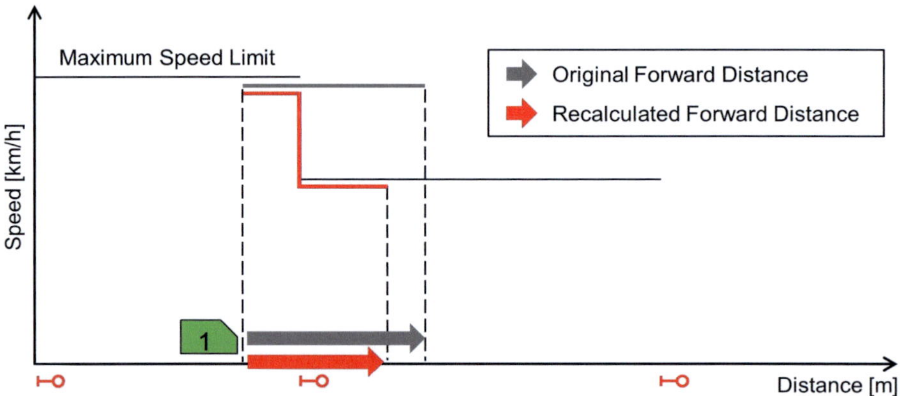

Figure 3-14: Update Train Head Position on the Mesoscopic Level

3.1.4 The Components and Workflow on the Macroscopic Level

On the macroscopic level, an infrastructure network can be very briefly depicted with nodes and links [Radtke, 2014]. Stations and junctions are abstracted into nodes, and open track sections between nodes are abstracted into links. In order to fulfill the requirement of dispatching tasks on the macroscopic level, the definitions of loop node, junction node and open track section are specified in [Cui, 2010]. To enable the infrastructure model be applicable for dispatching tasks in the multi-scale model, the macroscopic infrastructure hierarchy defined in [Cui, 2010] is modified and expanded in this approach. The new infrastructure hierarchy is shown in Figure 3-15.

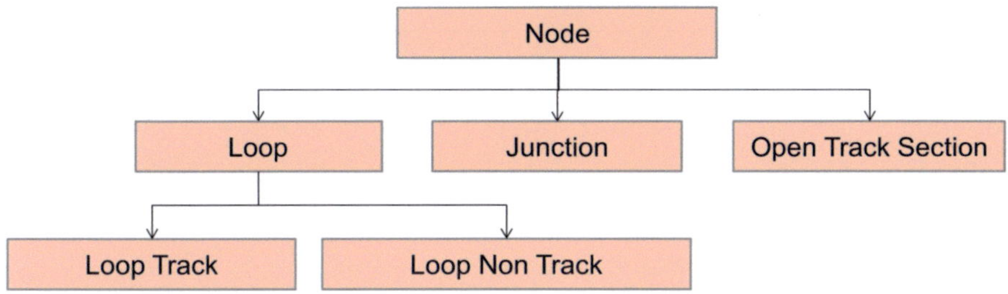

Figure 3-15: Class Diagram for Infrastructure Hierarchy

The same as the macroscopic infrastructure model defined in [Cui, 2010] three types of nodes are used herein: loop node, junction node and open track section. On the macroscopic level, these three types of nodes are enough to describe an infrastructure network. However, on the microscopic and mesoscopic levels, inside a loop node it is necessary to distinguish loop track and loop non track to fulfill the requirement of dispatching tasks (the applications of loop track and loop non track are to be elaborated in Section 5.2.2). A loop is defined as an operational site where the dispatching actions - overtaking and passing – are able to be performed. Loop tracks are the tracks on which scheduled or unscheduled stops of trains can be performed inside loop nodes. The other part of a loop node (except loop track) refers to the loop non track. A junction node is consisted of at least one junction-type resource, but overtaking and passing are not able to be performed in it. The infrastructures between nodes (i.e. loop node or junction node) are defined as open track sections[13].

Due to the high abstraction level of the macroscopic model, block sections do not exist anymore, which is the basics to ensure safe train separation on the microscopic and mesoscopic level. Therefore, trains have to be temporally separated on the macroscopic level. On open track sections, safe train separation is realized with minimum line headways (for the details of calculation method of minimum line headways it is

---

[13] The block sections consisted of exclusively free resources is treated as a special type of open track section. This kind of open track section will not be considered in the dispatching process.

referred to [Pachl, 2014]). Similar to the macroscopic model in [Cui, 2010], arrive-arrive headways (denoted by AA) are used to separate trains with successive movements, and depart-arrive headways (denoted by DA) are used to separate trains with opposite movements (Figure 3-16). In addition, the throughput is also a constraint on open track section. The throughput of a macroscopic node is defined as the maximum number of trains that can simultaneously occupy it. The throughput of an open track section is direction dependent. The throughput for a certain direction is set to the number of its included block sections in the same direction. Last but not least, sequence of trains cannot be changed on an open track section, so FIFO principle (first-in-first-out) must be followed.

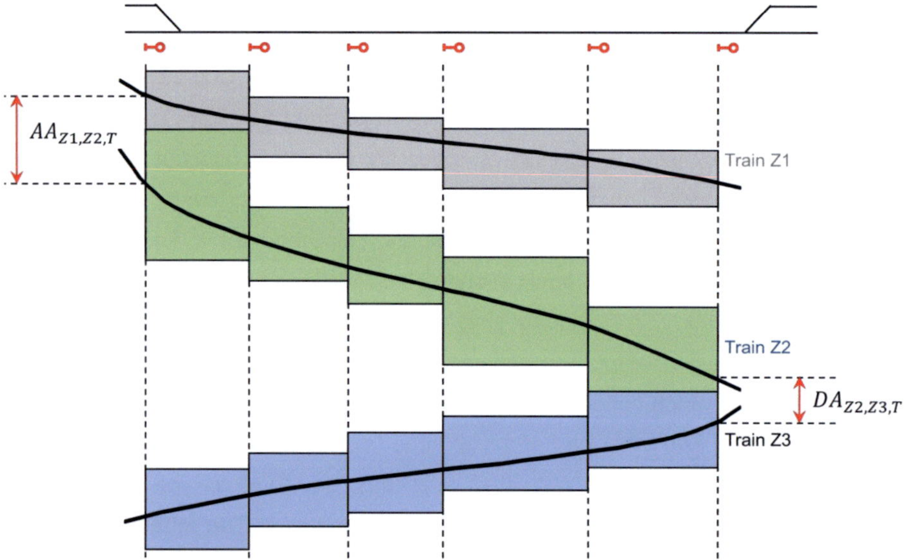

Figure 3-16: Minimum Line Headway on Open Track Section T (modified from [PT1, 2016] and [Cui, 2010])

Inside a loop node, the information on the included loop tracks are maintained, in order to ensure the consistency of the dispatching optimization algorithm on different description levels (see Chapter 5). The throughput of a loop track is defined as one. When a loop node is requested by a train, only the throughput of the corresponding loop track needs to be checked. Furthermore, train sequences can be altered in loop nodes, such as through overtaking. Therefore, train sequence will not be considered as a constraint for train movement in loop nodes.

To regulate train movement in junction nodes, both the throughput and train sequence should be considered. In a junction node, the relative sequence of trains coming from the same open track section cannot be altered, because it is impossible to carry out overtaking actions in junction nodes. So the FIFO principle is also adopted in junction nodes.

Train movements will be simulated with the assistance of these three kinds of macroscopic resources, and the departing/passing time on each macroscopic node is interested. On the macroscopic level, the whole path of a train is abstracted into a macro-path, and the current position of the train refers to this current macroscopic node. A macro-path is defined as a sequence of macroscopic nodes derived from a given timetable [Cui, 2010]. In the process of resource requirement, only the trains, which have complete their scheduled operation on their current nodes, have the chance to request new resources. The completeness of a scheduled operation can be checked as follows:

$$T_{now} \geq TB_{i[N]-1,Z_j} + TI_{i[N],Z_j} \tag{3-1}$$

Notation used (the notations are the same or similar to those used in [Cui, 2010]):

$i[N]$:   Index of the macroscopic node N in the macro-path of a corresponding train

Z_j: Train Z_j

T_{now}:   Current execution time in the simulation model

$TB_{i[N]-1,Z_j}$: Departing/passing time for train Z_j in the $(i[N]-1)^{th}$ node of its macro-path

$TI_{i[N],Z_j}$: Scheduled operational time for train Z_j in the $(i[N])^{th}$ node of its macro-path

The arrival time onto the current node is equal to the departure/passing time[14] from the last node ($TB_{i[N]-1,Z_j}$). Both the scheduled running time and scheduled dwell time on the current node are included in the scheduled operational time ($TI_{i[N],Z_j}$). Besides the operational time, the FIFO constraint should be taken into account if the current node of the considered train is a junction node or an open track section. Only if the train is the first occupier of the current node, the train can request new resources; otherwise the train is not allowed to leave the current node.

In the process of resource allocation, resource requirements ought to be checked with the conflict-free and deadlock-free test. For the conflict-free test, only throughput needs to be considered if the requested resource is a loop node or junction node. As long as the throughput of the requested node is not consumed yet, the resource requirement will be directly labeled as conflict-free. For open track sections, besides the throughput, conflict-free resource requirement also needs to fulfill the minimum line headway constraints (i.e. AA and DA). Depending on the running directions of the requester (denoted by Z_j) and its immediately previous train (denoted by Z_{Prev}) on the requested open track section (denoted by T), either AA (successive movement see Formula 3-2) or DA (opposite movement see Formula 3-3) should be accordingly fulfilled.

$$T_{now} \geq TB_{i[T]-1,Z_{Prev}} + AA_{Z_{Prev},Z_j,T} \tag{3-2}$$

$$T_{now} \geq TB_{i[T],Z_{Prev}} + DA_{Z_{Prev},Z_j,T} \tag{3-3}$$

Notation used:

$DA_{Z_{Prev},Z_j,T}$: Depart-arrive headway between the train Z_j and Z_{Prev} on open track section T

$AA_{Z_{Prev},Z_j,T}$: Arrive-arrive headway between the train Z_j and Z_{Prev} on open track section T

---

[14] The initial values of all departure/passing times are positive infinite. During the simulation process, only after a train has physically departed from a certain node, the corresponding departure/passing time could be updated to the actual value.

In the process of proceeding with simulation tasks, the procedures to be carried out are depending on the results of resource allocation. If no resource requirement occurred or requested resources were not allocated to the train, do nothing; otherwise the followings procedures should be carried out:

−   remove the train from the occupier list of the current node,

−   add the train into the occupier list of the next node,

−   record the departing/passing time on the current node,

−   update the current position of the train to its next node.

3.2   Downscaling Method for the Multiscale Model

In the top-down approach, the down-scaling method is designed to estimate the detailed infrastructure attributes on a lower level based on the abstracted attributes on an upper level. Because the detailed infrastructure attributes could be unknown, the result of down-scaling is an artificial infrastructure network. Whether the artificial infrastructure network is topologically similar to the actual infrastructure network is not of interest herein. The aim of the down-scaling method is to generate an artificial infrastructure network which gives similar simulation results as the actual one.

Given a macroscopic infrastructure network, an artificial microscopic infrastructure network can be generated with the macroscopic infrastructure manager of the software RailSys [RMCon, 2016; Gille et al., 2008]. Two important steps are included in the process:

−   Specify the accessibilities from platforms to open track sections and define route exclusions in stations (Figure 3-17)

−   Generate infrastructure elements such as signals, release contacts and block sections for both stations and open track sections (Figure 3-18)

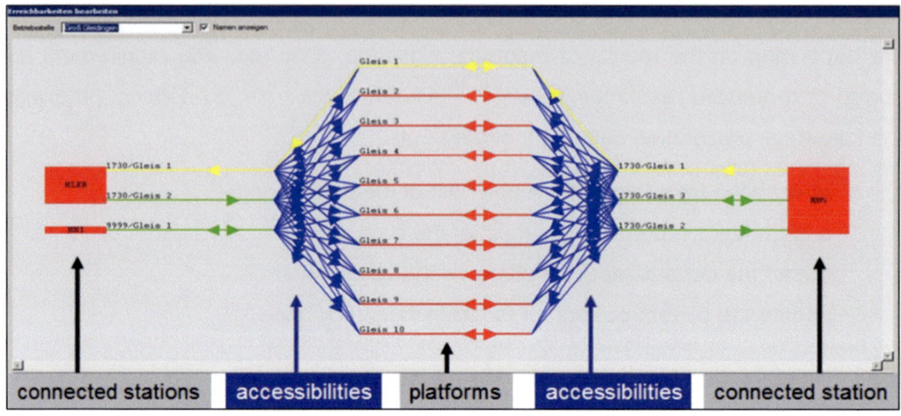

Figure 3-17: Accessibilities from Platforms and Open Track Sections [RMCon, 2016; Gille et al., 2008]

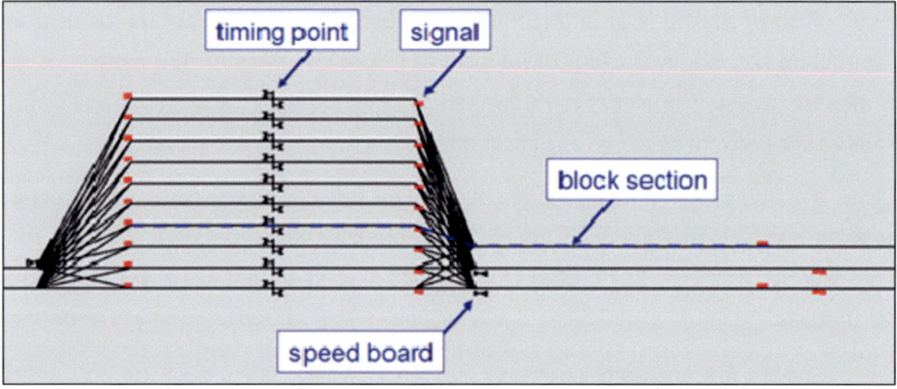

Figure 3-18: Generate Infrastructure Elements for Microscopic Infrastructure Network [RMCon, 2016; Gille et al., 2008]

Isometric block sections can be used (the lengths of block sections are kept the same), and accordingly the length of the edges can be derived. As seen in Figure 3-18, a customized standardized mesoscopic infrastructure network is eventually generated based on the rationalities in railway construction and operation. The algorithm of generating a standardized microscopic infrastructure network is not covered within this approach, and the details of the algorithm can be observed in [Gille et al., 2008].

Before the artificial microscopic infrastructure network can be put into use, the other attributes of the edges - permissible speed, gradient and radius - should be specified. There might be already a significant difference between the artificial and the actual infrastructure network, so only the most significant attribute – permissible speed - are taken into account, rather than the less significant attributes – gradient and radius[15].

The setting of permissible speed has a direct influence on the calculation of operational time, which is a basic function of a simulation model. The actual operational times of a train in the infrastructure nodes along its path are given variables in the macroscopic simulation model (see Section 3.1.4). The operational time calculated based on the artificial network should approximate these true values as much as possible. So, total variation of operational times is chosen as the indicator to evaluate the quality of the artificial infrastructure network. Since the operational times are designated separately to each infrastructure node, the qualities of the infrastructure nodes are also separately evaluated. This is intended to optimize the quality of an artificial infrastructure node by adjusting the permissible speeds of its included edges:

$$\sum_{j=1}^{j=n_{ges}} \left| TI_{i[N],Z_j}^{ART} - TI_{i[N],Z_j} \right| \to min \tag{3-4}$$

Notation used:

$i[N]$:   Index of the macroscopic node N in the macro-path of a corresponding train

Z_j: Train Z_j

$TI_{i[N],Z_j}$:   Real scheduled operational time for train Z_j in the $(i[N])^{th}$ node of its macro-path

$TI_{i[N],Z_j}^{ART}$:   Artificial scheduled operational time for train Z_j in the $(i[N])^{th}$ node of its macro-path

n_{ges}:   Total number of trains running through the infrastructure node N

---

[15] The values of gradient and radius of each edge should be set to zero.

Theoretically, every individual edge in an infrastructure node may have a different permissible speed. However, the data given is currently insufficient to infer the permissible speed of each individual edge[16]. So in an infrastructure node a common permissible speed is designated to all the edges in the same direction. The common permissible speed in each direction should be adjusted until the minimum of the total variation of operational times is reached.

Take a simple open track section consisted of five isometric block sections as an example[17]: the length of every block section is 1500 meter, and the original permissible speeds of the edges constituting the block sections are marked in blue in Figure 3-19. Three trains, which are denoted as Z1 (long distance passenger train), Z2 (short distance passenger train) and Z3 (freight train), are scheduled to run through this open track section. The maximum speeds of them are 200, 120 and 100 km/h, respectively. The real operational times of these three trains on this open track section were calculated with the microscopic model, and the results are shown in Table 3-1.

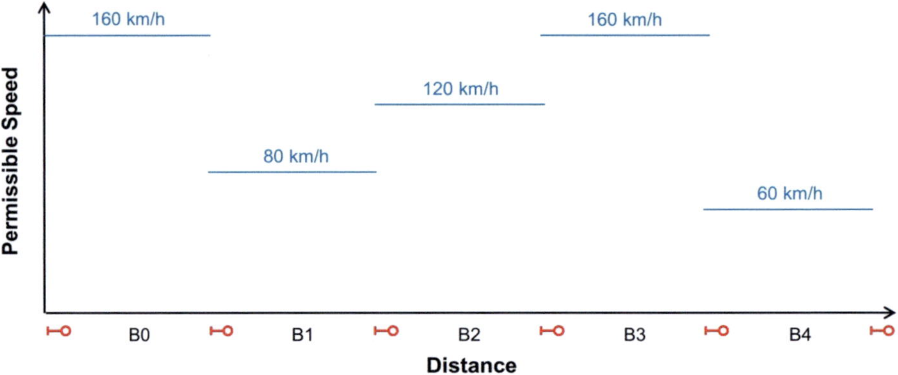

Figure 3-19: Original Open Track Section

In the downscaling process, the common permissible speed of all edges was adjusted, and the corresponding total variation of operational times was calculated with Formula (3-4). The results of the adjustment are illustrated in Figure 3-20.

---

[16] The number of edges in an infrastructure node (see page 35) can be far larger than the number of trains running through the infrastructure node.

[17] In this example, the original and the generated artificial infrastructure network are kept the same, in order to solely evaluate the influence of permissible speed.

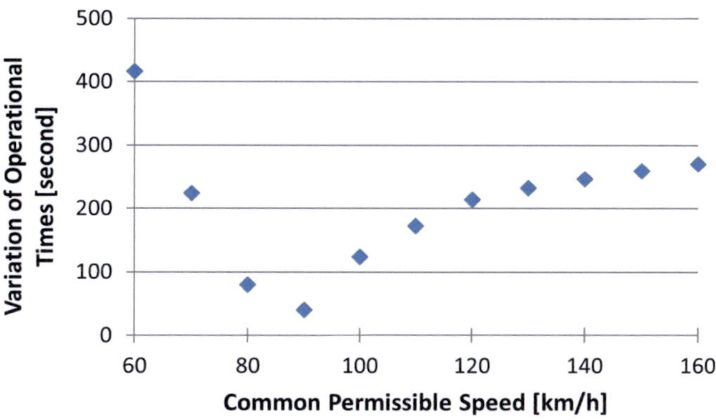

Figure 3-20: Adjustment of Common Permissible Speed for Downscaling

The optimal common permissible speed is 90 km/h, and the minimum total variation of the operational times is 40 seconds. Obviously, the total variation of operational times on condition of optimal common permissible speed is significantly smaller than the deviations on condition of the other speeds. Therefore, to ensure the accuracy of the artificial infrastructure network, optimization of common permissible speed is indispensable in the down-scaling process. In other words, optimization of common permissible speed could increase the accuracy of the generated artificial infrastructure network significantly. In order to analyze the variation in detail, the operational time of each train calculated based on the artificial infrastructure network is listed in the following table.

	TI_{Z1}[second]	TI_{Z2}[second]	TI_{Z3}[second]
Original Infrastructure	298	305	334
Optimal Artificial Infrastructure	301	301	301
Individual Variation	3	4	33

Table 3-1: Operational times on Condition of Original or Optimal Artificial Infrastructure

It can be seen that the individual variations for the train Z1 and Z2 are negligible (circa 1%), but the individual variation for the train Z3 is substantial (circa 10%). Furthermore, with the common permissible speed of 90km/h, the operational times of these three trains become identical, which weakens the differences of the different train types. Even through the overall quality of the artificial infrastructure network can be optimized with the method proposed in this section, certain substantial variations of operational times and convergence of operational times may still lead to inaccurate simulation results.

In summary, compared to the macroscopic infrastructure, the advantage of the generated artificial microscopic infrastructure is that a higher level of detail of the infrastructure network could be depicted, such as platforms in stations, accessibilities between platforms and open track sections and exclusive station routes. The artificial microscopic infrastructure can be generated with the macroscopic infrastructure manager of the software RailSys, and the permissible speeds of edges can be optimized with the downscaling method developed in this section. The downscaling method is only recommended, when detailed simulation results are required but no detailed and consistent infrastructure data is available. For applications with a high demand on accurate running time calculation, such as dispatching and timetabling tasks, the original detailed infrastructure data should be used, since the artificially generated data can hardly meet such demands.

4 Assessment Method for the Multi-scale Model

The computation complexity and the simulation accuracy should be balanced when working with a large investigation area. It can be achieved by using the multi-scale mode developed in Chapter 3. The significance value determines the appropriate description level of each infrastructure node (i.e. loop node, junction node and open track section). The higher the significant value of an infrastructure node, the more likely the node is to be described in greater detail. The significance value contains two indicators: the relevance to conflicts, which have a positive correlation with the significance value, and the aggregation accuracy, which have a negative correlation.

4.1 Calculation Method of Significance Value

In order to determine the relevance of an infrastructure node to conflicts, the delay propagation scope should be estimated at first. The infrastructure nodes inside the delay propagation scope will be given a higher degree of correlation compared to the nodes outside of the scope. Take advantage of the existing simulation model (i.e. the multi-scale model), the information on conflicts, such as the locations of conflicted trains and the durations of hindrances, can be accurately recorded during the simulation process. Thus, the delay propagation scope can be deduced based on this information. The estimation method of delay propagation scope will be hereinafter explained based on a small example.

As the result of a conflict, the hindered train will experience a knock-on delay on a certain block. From this moment on, the delay will accompany the train along its further path until the elimination of the delay or the termination of the train run. Take the same case in Figure 4-1 as an example. The trains Z1 and Z2 are located on the block section B0 and B1, respectively, and Z2 is hindered by Z1 by two minutes. This directly results in the knock-on delay of Z2 on B0 (denoted by $tw_{2,0}$). Subsequently $tw_{2,0}$ will propagate in the form of delay (denoted by $td_{2,i}$) along the further path. Using the recovery time $td_{2,i}$ is eventually eliminated on B4. So the area from B0 to B4 is taken as the delay propagation scope for this example.

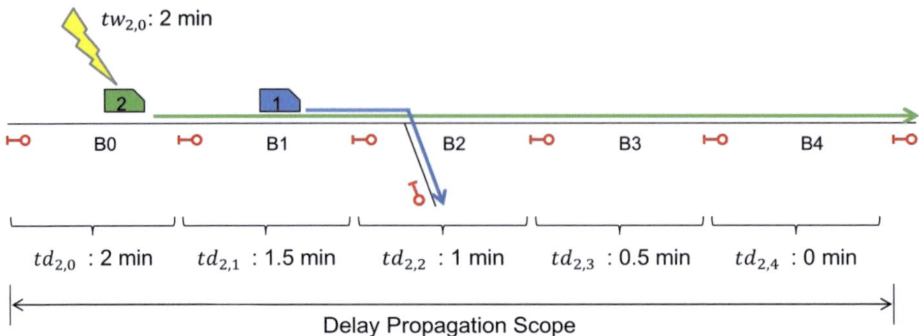

Figure 4-1: An Example of Delay Propagation Scope

Before $td_{2,i}$ is eliminated, Z2 may hinder the other trains. The newly generated delays of the affected trains will propagate along their own paths, and the delay propagation scope is consequently enlarged. It can be seen that the union of the delay propagation scopes of all affected trains is the overall delay propagation scope. So, once a train is hindered in the prediction period, it will be traced on its further path until the delay is eliminated or the train run or the prediction time is terminated. In case the train is hindered again by another train in between, it is necessary to update the delay of the hindered train and continue with tracing. The union of the delay propagation of all hindered trains will be outputted as the overall delay propagation scope[18] at last.

Accordingly, the infrastructure nodes can be divided into two sets (i.e. inside or outside of the scope), and the infrastructure nodes outside of the delay propagation scope are assigned higher priority to be aggregated compared to their counterpart. Moreover, the infrastructure nodes in the same set should be further ranked based on their aggregation accuracy, and the infrastructure nodes with higher aggregation accuracy should be assigned higher aggregation priority. Both for dispatching objective function and for capacity research, knock-on delay is an important variable, so

---

[18] Dispatching decisions can change the conflict relationship between trains in the simulation process. However, the final dispatching decisions are still unknown at the moment when the overall delay propagation scope should be determined. Therefore, the simplest dispatching principle FCFS (First Come First Serve) is adopted for the estimation of delay propagation scope and the result can be taken as the upper bound of delay propagation scope.

the variation of total knock-on delay is chosen as the indicator of aggregation accuracy.

Through the use of the aforementioned approach, a controlled experiment is set up in order to evaluate the variation in the values of total knock-on delay through the aggregation of an infrastructure node. This experiment involves the separation of search subjects into two groups, the control group and the experimental group. A certain investigated variable is changed in the experimental group, and all variables are kept constant in the control group. A comparison of the results of the two groups displays a quantified measure of the effects of the variation of the investigated variable. In order to determine aggregation accuracy, all infrastructure nodes are maintained on a microscopic level apart from the investigated node, which is abstracted onto the mesoscopic level. Meanwhile, both the timetable as well as the stochastic deviations during the operation process is kept identical for both infrastructure scenarios. Both scenarios must undergo a number of operational simulations, and the average total knock-on delays are calculated based on the simulation protocol. The aggregation accuracy of the investigated infrastructure node is defined as the absolute value of the difference between the average total knock-on delays of the two infrastructure scenarios (Formula 4-1).

$$ACC_N^{micro \rightarrow meso} = \left| \sum_{j=1}^{n_{ges}} \sum_{i=1}^{i=Z_j} tw_{j,i}^{CG} - \sum_{j=1}^{n_{ges}} \sum_{i=1}^{i=Z_j} tw_{j,i}^{EG} \right| \tag{4-1}$$

Notation used:

$ACC_N^{micro \rightarrow meso}$: Aggregation accuracy of infrastructure node N from the microscopic level to the most detailed mesoscopic level

$tw_{j,i}^{CG}$, $tw_{j,i}^{EG}$: Knock-on delay of train j on block section i in the controlled group (CG) or experimental group (EG)

n_{ges}: Total number of trains

zj: Amount of block sections along the path of train j

The controlled experiment outlined above may be used to precisely calculate the aggregation accuracy of abstracting any infrastructure node from the microscopic to a concrete mesoscopic level. However, a large amount of variants of the mesoscopic infrastructure network are resulted on the mesoscopic level of the multi-scale model, due to it being characterized by continuous scaling. The combination of any two adjacent occupation units will result in the generation of a new variant of the mesoscopic infrastructure network. If, in this case, the aggregation accuracy of continuous scaling on the mesoscopic level is evaluated further with the controlled experiment, the computational effort due to the increased variants will become infeasible. The loss of accuracy when aggregating from the most detailed mesoscopic level to a more abstracted mesoscopic level is limited when compared to that caused by the aggregation from the microscopic level to the most detailed mesoscopic level. Aggregation from the microscopic to the most detailed mesoscopic level (see Section 3.1.3), results in the simplification of both train movement behavior and regulation of signaling systems. In the case of aggregation from the most detailed mesoscopic level to a more abstracted mesoscopic level (see Section 3.1.3), the main changes involve the releasing times of the involved occupation units, due to the combination of adjacent occupation units. The conclusion can therefore be made that the controlled experiment should be restricted to the calculation of the aggregation accuracy of an infrastructure node abstracted from the microscopic level to the most detailed mesoscopic level, and that a simpler and more computationally efficient method should be developed to estimate the aggregation accuracy of the continuous scaling on the mesoscopic level, which will be elaborated separately in Section 4.2. A crucial output of the method is thus returned in the form of the aggregation accuracy of combining two adjacent occupation units (Formula 4-2).

$$ACC_{R_k+R_h}^{meso} = |\Delta E(T^{OVLP,R_k+R_h})| \tag{4-2}$$

Notation used:

ACC_{R_k,R_h}^{meso}: Aggregation accuracy of two occupation units R_k and R_h on the mesoscopic level

$\Delta E(T^{OVLP,R_k+R_h})$: Relative change of expected total overlapping time period caused by combination of R_k and R_h (details see Section 4.2)

The following formulas outline the calculation of the significance value of each infrastructure node (abstracted from the microscopic to the most detailed mesoscopic level), as well as the significance value of a larger mesoscopic occupation unit comprised of two other occupation units (abstracted from a certain mesoscopic level to a more abstracted mesoscopic level).

$$SV_N^{micro \rightarrow meso} = \chi_{\{l|l \in S_p\}}(N) \cdot M - ACC_N^{micro \rightarrow meso} \tag{4-3}$$

$$SV_{R_k+R_h}^{meso} = \chi_{\{l|l \in S_p\}}(N) \cdot M - ACC_{R_k+R_h}^{meso} \tag{4-4}$$

Notation Used:

$SV_N^{micro \rightarrow meso}$: Significance value of an infrastructure node N to be abstracted from the microscopic level to the most detailed mesoscopic level

SV_{R_k,R_h}^{meso}: Significance value of a large mesoscopic occupation unit composed of R_k and R_h on the mesoscopic level

$\chi_{\{l|l \in S_p\}}(N)$: Indicator function, if the infrastructure node N belongs to the delay propagation scope S_p , it is equal to 1, otherwise 0.

M: A sufficiently large number, which is at least larger than the maximum of the aggregation accuracies

The significance values form the basis of a two-step process in the establishment of the description levels of infrastructure nodes. The first step consists of the sorting of the infrastructure nodes according to their significance values (Formula 4-3). Computational capacities, time constraints and other on-site factors can limit the optimal number of occupation units on each description level. This is not within the scope of this research project. For the sake of explanation, assumed values will be utilized within this approach. The optimal amount of occupation units on each description level and the ranking of the significance values of the infrastructure nodes determine the description level of each infrastructure node (Figure 4-2).

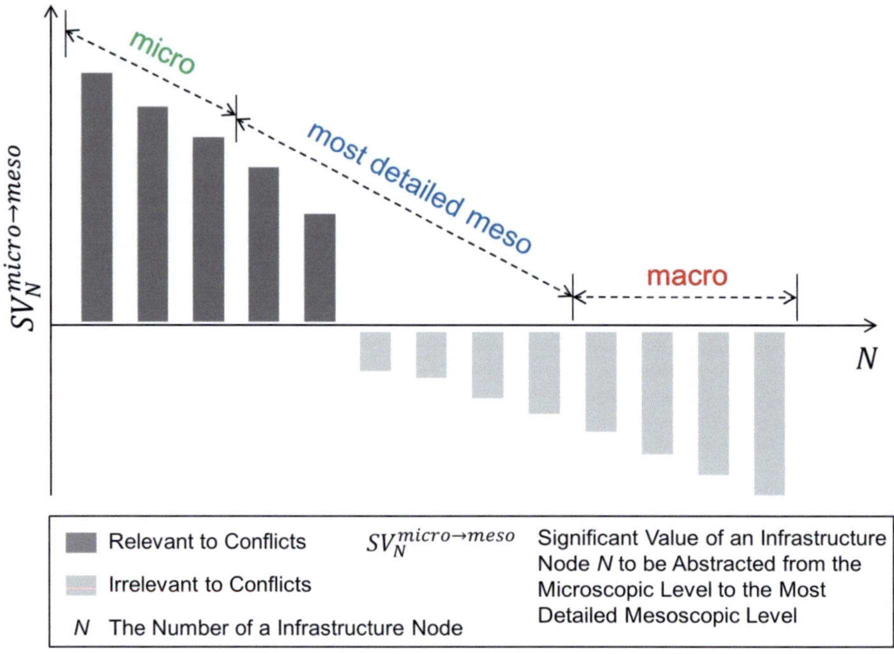

Figure 4-2: Determination of the Description Level of Each Infrastructure Node (I)

The second step of the process aims at allowing less significant nodes to be simulated on the mesoscopic level, rather than the macroscopic level, through further aggregation of the occupation units on the mesoscopic level. This is carried out by the aggregation of the occupation units in sequence according to their significance values calculated with Formula (4-4). The aggregation process terminates at the moment at which no further infrastructure nodes are able to be transformed from the macroscopic to the mesoscopic level. The final result obtained is the exact number of nodes on each description level, as well as their specific forms (Figure 4-3).

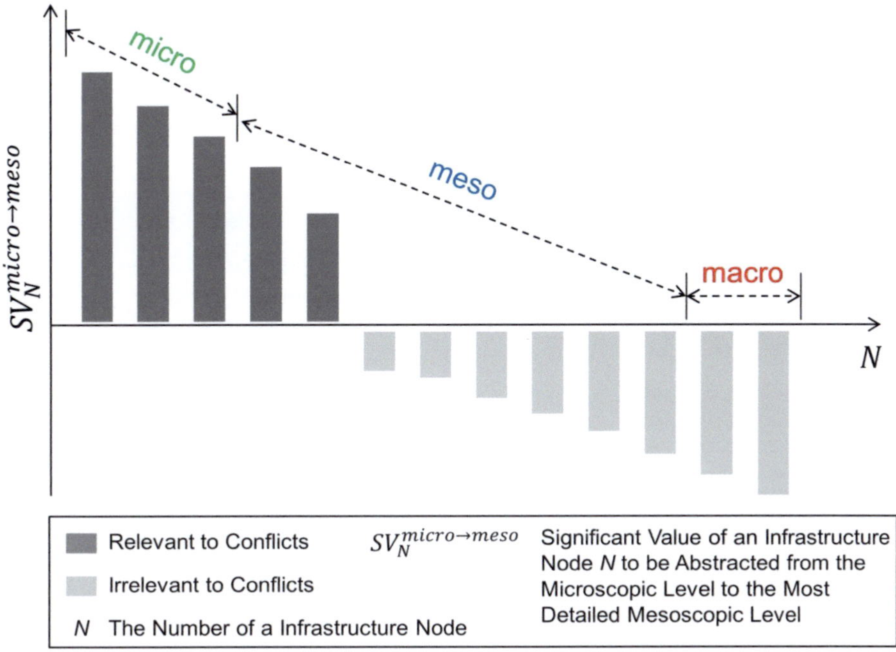

Figure 4-3: Determination of the Description Level of Each Infrastructure Node (II)

4.2   Further Aggregation on Mesoscopic Level

The further aggregation of basic structures on the mesoscopic level is carried out in order to facilitate the simulation of a larger number of infrastructure nodes on the mesoscopic level. This aggregation causes a variation in the blocking times of train runs on the concerned basic structures and can, therefore, have an influence on the conflicts between trains. An example of this phenomenon is shown in Figure 4-4, in which two basic structures, with two train runs arranged on them, are to be aggregated. The conflict situation is presented as an overlapping of the blocking times of the concerned trains before aggregation. After the aggregation process, the blocking time of Train Run 1 is adjusted, due to the absence of partial releasing. The conflict is either enlarged (throughput = 1) or disregarded (throughput = 2) according to the throughput of the new occupation unit.

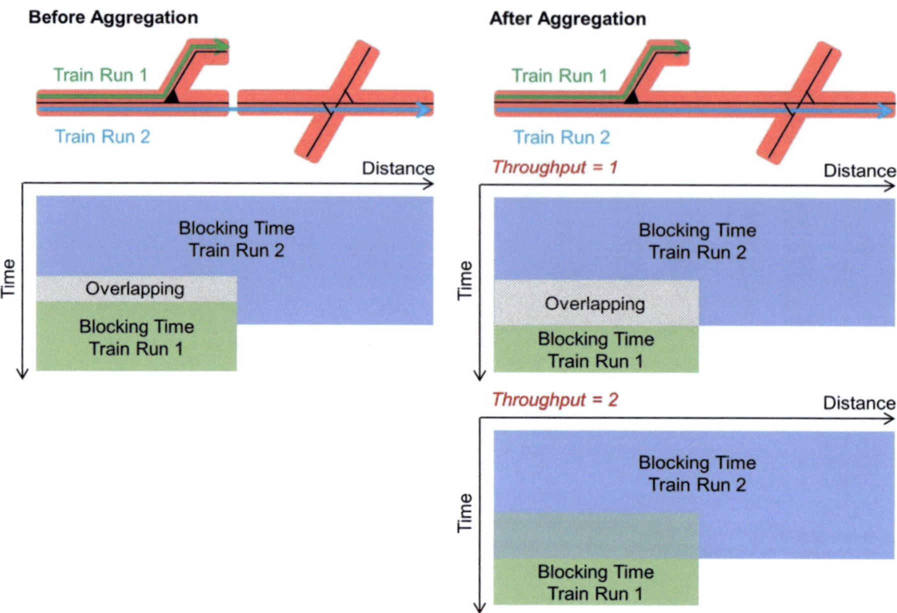

Figure 4-4: Comparison of Overlapping of Blocking Times Before and After Aggregation

From the aforementioned description, the relative change of conflicts between train runs before and after aggregation, also seen as the relative change of overlapping of blocking times, can be established as a good indicator in quantifying aggregation accuracy. This relative change of conflicts may influence further train runs as it propagates, which can be accurately estimated with the method outlined in Section 4.1. However, due to the large amount of aggregation possibilities even on a small infrastructure network, this may result in infeasible computational requirements. In practical terms, the relative change of conflicts will therefore be minimized from the source, and its further influences will be ignored.

Two variables are used in the description of the blocking time of a train run on a certain occupation unit: start blocking time and the length of blocking time. It is possible that the blocking time is extended in case of hindrances (e.g. unscheduled stop at a red signal). The average extension of blocking time can be determined through a large amount of operational simulations. The expected length of the actual blocking time is the summation of the length of the scheduled blocking time and its average extension, and it is used in the estimation of overlapping times. Within the following examples, unless otherwise stated, the length of a blocking time refers specifically to the expected length of the actual blocking time. Taking into account deviations in actual operation, the actual start blocking time is expressed as:

$$t_{j,i}^{start,Ist} = t_{j,i}^{start,Soll} + td_{j,i} \qquad (4\text{-}5)$$

Notation used:

$t_{j,i}^{start,Ist}$:   Actual start blocking time of train j on occupation unit i

$t_{j,i}^{start,Soll}$:   Scheduled start blocking time of train j on occupation unit i

$td_{j,i}$:   Delay of train j on occupation unit i

It can generally be assumed that $td_{j,i}$ follows Erlang-K distribution. In this research negative exponential distribution was used, which is a special case of Erlang-K distribution. Thus, $t_{j,i}^{start,Ist}$ follows the same probability distribution, because $t_{j,i}^{start,Soll}$ is a constant value fixed by the train schedule.

Discrete time is considered rather than continuous time for the sake of convenience in computerized processing. This selection requires that all time-related variables being considered, including the start and end time of the investigated time period, the length of blocking time and the start blocking time, be adjusted to become integral multiples of the discrete interval. Furthermore, a frequency histogram can be utilized to express the probability distribution of delay using the same time interval (e.g. the left subgraph in Figure 4-5).

The actual blocking time occurs at different time periods with a certain probability depending on a variety of possible delays. An example of this is shown in the right subgraph of Figure 4-5: for a certain delay x ($x \in \{0, \Delta t, 2\Delta t, 3\Delta t\}$) the start blocking time $(x + t_s)$ can be accordingly determined, and the time intervals within the length of blocking time from $(x + t_s)$ to $(x + t_s + 3\Delta t)$ will be blocked. The probability of the blocking time occurring in this time period is equal to the probability of the occurrence of the delay x. Due to the fact that time is discretized, the probability that delay x will occur is defined as the probability that the delay belongs to the time interval [x, x+Δt).

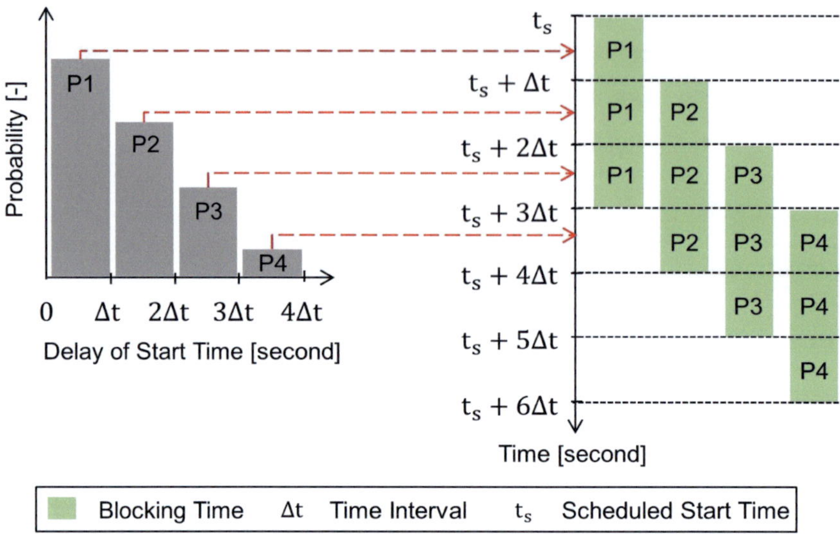

Figure 4-5: Probability Distribution of Delay and Blocking Time

The blocking probability of each time interval can be further determined by transforming the probability distribution of the blocking time, as shown in Figure 4-6. In this example, the blocking probabilities of the time intervals outside of the range $[t_s, t_s + 6\Delta t)$ should be set to zero.

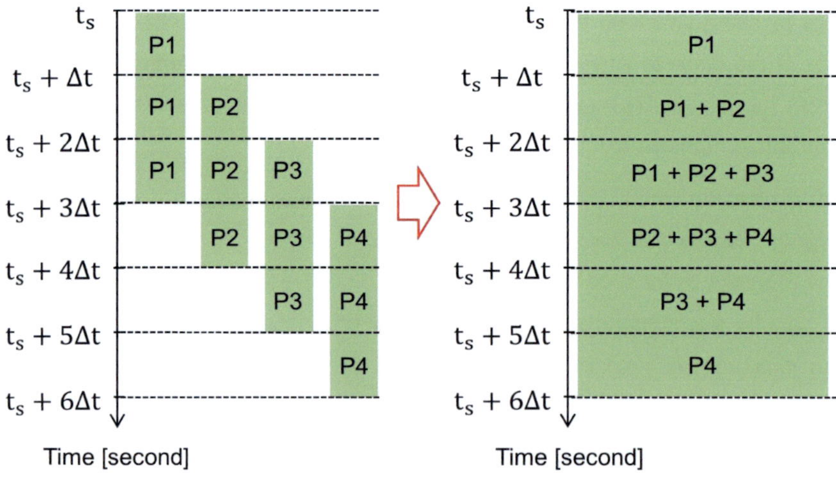

Figure 4-6: Blocking Probability of Each Time Interval

Through the method above, the blocking probability of each time interval can be determined for each involved train run on a given occupation unit, and thereby the expected overlapping of blocking times between train runs in each time interval can be calculated. The sum of these results comprises the expected total overlapping time in the entire investigated time period, and can be used for the calculation of the relative changes of conflicts before and after aggregation. It can be seen that the most crucial factor in the method is to calculate the expected overlapping in each time interval. A basic example is detailed in Figure 4-7 in order to facilitate understanding of the calculation method. Within the example three train runs, Z1, Z2 and Z3, are considered on an occupation unit named R3, and the investigated time period is from t_0 to t_m. The blocking probability of each time interval for each train run is given in the second subgraph, and $[t_2, t_3)$ is chosen as a sample time interval.

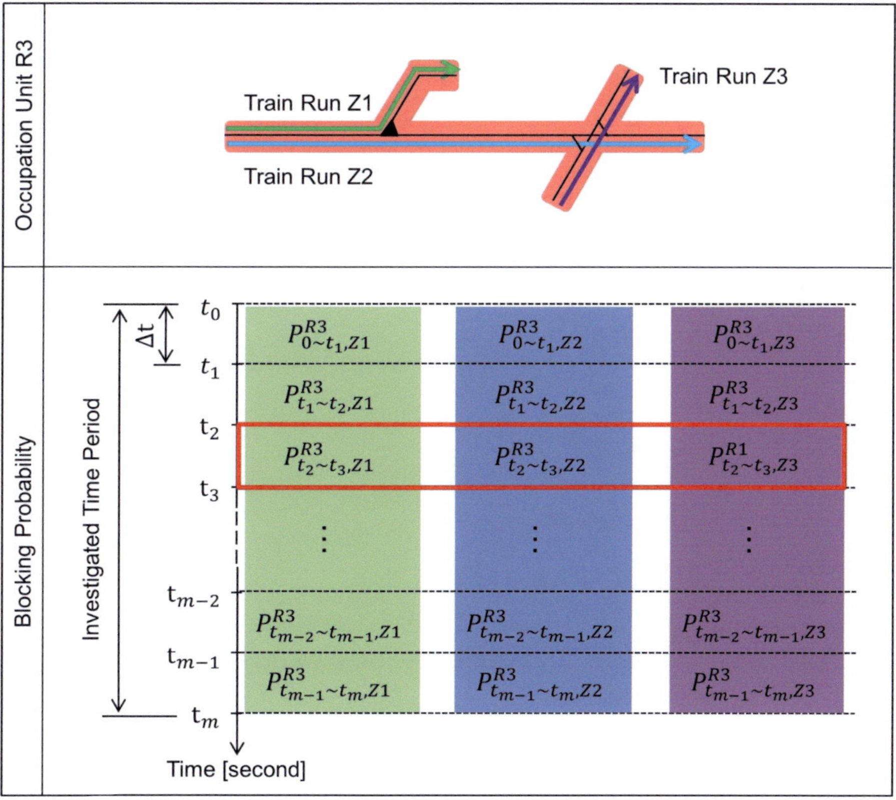

Figure 4-7: Calculation of Expected Overlapping Time Period on an Occupation Unit

It is evident that, on this occupation unit, conflicts can occur between every two trains or between all three trains. So, there are $(C_3^2 + C_3^3)$ possible conflict situations[19]. Furthermore, because R3 is composed of two basic structures (see Figure 4-4), the throughput[20] of R3 is either 1 or 2.

Take Z1 and Z2 as an example of the conflict situation occurred only between two trains, and the expected overlapping time period between them in the time interval $[t_2, t_3)$ can be calculated with the general formula (4-6).

$$E(T_{t_2 \sim t_3, Z1, Z2}^{OVLP, R3}) = P_{t_2 \sim t_3, Z1, Z2}^{R3} \cdot t_{t_2 \sim t_3, Z1, Z2}^{OVLP, R3} \tag{4-6}$$

Notation used:

$P_{t_2 \sim t_3, Z1, Z2}^{R3}$:   Occurrence probability of the conflict situation between Z1 and Z2 on the occupation unit R3 in the time interval from t_2 to t_3

$t_{t_2 \sim t_3, Z1, Z2}^{OVLP, R3}$:   Total overlapping time period between Z1 and Z2

$E(T_{t_2 \sim t_3, Z1, Z2}^{OVLP, R3})$:   Expected total overlapping time period between Z1 and Z2

The occurrence probability of the corresponding conflict situation can be easily calculated based on the blocking probability of each time interval (e.g. second subgraph in Figure 4-7) as follows:

$$P_{t_2 \sim t_3, Z1, Z2}^{R3} = P_{t_2 \sim t_3, Z1}^{R3} \cdot P_{t_2 \sim t_3, Z2}^{R3} \cdot (1 - P_{t_2 \sim t_3, Z3}^{R3}) \tag{4-7}$$

Where:

$P_{t_2 \sim t_3, Zj}^{R3}$:   Blocking probability of the time interval $[t_2, t_3)$ for Zj on the occupation unit R3

---

[19] If n train runs exist on a certain occupation unit, the amount of possible conflict situations are $(C_n^2 + C_n^3 + \cdots + C_n^n)$. To calculate the overlapping time period in a certain time interval, all possible conflict situations should be considered.

[20] In general, a larger occupation unit (designated as R3) is composed of another two occupation units (designated as R1 and R2). The throughputs of R1 and R2 are TP_{R1} and TP_{R2} respectively. The range of TP_{R3} is $[1, TP_{R1} + TP_{R2}]$

The total overlapping time period in the corresponding conflict situation changes as the throughput varies. In case that the throughput is 1, the total overlapping time period between Z1 and Z2 is

$$t^{OVLP,R3}_{t_2\sim t_3,Z1,Z2}\big|_{TP_{R3}=1} = \Delta t \tag{4-8}$$

In case that the throughput is 2, the overlapping between these two trains will be ignored.

$$t^{OVLP,R3}_{t_2\sim t_3,Z1,Z2}\big|_{TP_{R3}=2} = 0 \tag{4-9}$$

The expected overlapping time between Z1 and Z2 can be calculated:

$$E\big(T^{OVLP,R3}_{t_2\sim t_3,Z1,Z2}\big)\big|_{TP_{R3}=1} = P^{R3}_{t_2\sim t_3,Z1} \cdot P^{R3}_{t_2\sim t_3,Z2} \cdot \big(1 - P^{R3}_{t_2\sim t_3,Z3}\big) \cdot \Delta t \tag{4-10}$$

$$E\big(T^{OVLP,R3}_{t_2\sim t_3,Z1,Z2}\big)\big|_{TP_{R3}=2} = P^{R3}_{t_2\sim t_3,Z1} \cdot P^{R3}_{t_2\sim t_3,Z2} \cdot \big(1 - P^{R3}_{t_2\sim t_3,Z3}\big) \cdot 0 \tag{4-11}$$

On condition that conflicts occurred between these three trains, the occurrence probability of this conflict situation is:

$$P^{R3}_{t_2\sim t_3,Z1,Z2,Z3} = P^{R3}_{t_2\sim t_3,Z1} \cdot P^{R3}_{t_2\sim t_3,Z2} \cdot P^{R3}_{t_2\sim t_3,Z3} \tag{4-12}$$

In case the throughput is 1, the blocking times of every two of these three trains overlapped (there are C_3^2 possibilities). So the total overlapping time period is:

$$t^{OVLP,R3}_{t_2\sim t_3,Z1,Z2,Z3}\big|_{TP_{R3}=1} = C_3^2 \cdot \Delta t = 3 \cdot \Delta t \tag{4-13}$$

The throughput indicates the number of trains that can block an occupation unit without conflict. So the overlapping time periods occurred between the trains belonging to the throughout can be ignored[21]. When the throughput is 2, the total overlapping time period should be adjusted as follows:

$$t^{OVLP,R3}_{t_2\sim t_3,Z1,Z2,Z3}\big|_{TP_{R3}=2} = \big(C_3^2 - C_2^2\big) \cdot \Delta t = 2 \cdot \Delta t \tag{4-14}$$

---

[21] Generally speaking, if n_{cfl} trains are involved in a certain conflict situation and the throughput is n_{TP}, the number of overlapping is max $\{0,(C^2_{n_{cfl}} - C^2_{n_{TP}})\}$. n_{cfl} is always equal or larger than 2, so $C^2_{n_{cfl}}$ is equal or larger than 1. However, n_{TP} can equal to 1. In this case C_1^2 is equal to 0 by definition.

The expected overlapping times among Z1, Z2 and Z3 are:

$$E(T_{t_2 \sim t_3, Z1, Z2, Z3}^{OVLP,R3})|_{TP_{R3}=1} = P_{t_2 \sim t_3, Z1}^{R3} \cdot P_{t_2 \sim t_3, Z2}^{R3} \cdot P_{t_2 \sim t_3, Z3}^{R3} \cdot 3 \cdot \Delta t \qquad (4\text{-}15)$$

$$E(T_{t_2 \sim t_3, Z1, Z2, Z3}^{OVLP,R3})|_{TP_{R3}=2} = P_{t_2 \sim t_3, Z1}^{R3} \cdot P_{t_2 \sim t_3, Z2}^{R3} \cdot P_{t_2 \sim t_3, Z3}^{R3} \cdot 2 \cdot \Delta t \qquad (4\text{-}16)$$

The expected total overlapping time period in the time interval $[t_2, t_3)$ is the sum of the results of all possible conflict situations.

$$\begin{aligned}
E(T_{t_2 \sim t_3}^{OVLP,R3})|_{TP_{R3}=1 \, or \, 2} = \; & E(T_{t_2 \sim t_3, Z1, Z2}^{OVLP,R3})|_{TP_{R3}=1 \, or \, 2} \\[6pt]
& + E(T_{t_2 \sim t_3, Z2, Z3}^{OVLP,R3})|_{TP_{R3}=1 \, or \, 2} \\[6pt]
& + E(T_{t_2 \sim t_3, Z1, Z3}^{OVLP,R3})|_{TP_{R3}=1 \, or \, 2} \\[6pt]
& + E(T_{t_2 \sim t_3, Z1, Z2, Z3}^{OVLP,R3})|_{TP_{R3}=1 \, or \, 2}
\end{aligned} \qquad (4\text{-}17)$$

The same process as stated above can be used in the calculation of the expected total overlapping time periods in the other time intervals. Thereafter, the expected total overlapping time period in the entire investigated time period can be calculated as the sum of the results over all time intervals.

$$\begin{aligned}
E(T^{OVLP,R3})|_{TP_{R3}=1 \, or \, 2} = \; & E(T_{t_0 \sim t_1}^{OVLP,R3})|_{TP_{R3}=1 \, or \, 2} \\[6pt]
& + E(T_{t_1 \sim t_2}^{OVLP,R3})|_{TP_{R3}=1 \, or \, 2} \\[6pt]
& \;\; \vdots \\[6pt]
& + E(T_{t_{m-1} \sim t_m}^{OVLP,R3})|_{TP_{R3}=1 \, or \, 2}
\end{aligned} \qquad (4\text{-}18)$$

For the estimation of the aggregation accuracy for the two adjacent occupation units R1 and R2, it is required to determine the expected total overlapping time periods on them before aggregation with the same process in advance. The only difference is that the throughputs of R1 and R2 are pre-given. The relative change of expected total overlapping time period can be calculated as follows.

$$\Delta E\left(T^{OVLP,R3}\right)\big|_{TP_{R3}=1\ or\ 2} = \ E\left(T^{OVLP,R1}\right)\big|_{TP_{R3}=1\ or\ 2}$$

$$+E\left(T^{OVLP,R2}\right)\big|_{TP_{R3}=1\ or\ 2} \qquad (4\text{-}19)$$

$$-E\left(T^{OVLP,R3}\right)\big|_{TP_{R3}=1\ or\ 2}$$

The aggregation accuracy of R3 will then be indicated by comparing the absolute values of the relative changes in case of the different throughputs and selecting the lowest absolute value, and the corresponding throughput will be used as the representative throughput of R3.

An outline of the procedure is summarized below, and can be utilized to estimate the aggregation accuracy of further larger occupation units as well.

1) Calculate the expected overlapping time periods on two to be aggregated occupation units;

2) Adjust the blocking times of the concerned trains on the larger aggregated occupation unit and recalculate the blocking probabilities of each time interval for the trains;

3) Calculate the expected overlapping time period on the larger aggregated occupation unit;

4) Calculate the relative change of overlapping time period before and after aggregation and determine the aggregation accuracy and representative throughput for the larger occupation unit.

Those occupation units with higher aggregation accuracy, resulting from lower relative changes in expected overlapping time periods, will be granted higher aggregation priority in the upscaling process.

5 State-dependent Dispatching Algorithm

The multi-scale simulation model described in Chapter 3 is capable to generate a deadlock-free timetable implicitly following the simplest dispatching principle - First Come First Served (FCFS). With dispatching optimization algorithms, this basic timetable can be further improved. Many heuristic or metaheuristic algorithms are available to be used as the basis of the dispatching algorithm in railway operation. A widely used heuristic algorithm – greedy algorithm – is preferred in this approach.

On the one hand, the local search mechanism of the greedy algorithm is the basis of many heuristic and metaheuristic algorithms. Therefore, the specific local search mechanism developed in this section can be easily implemented under the framework of other heuristic or metaheuristic algorithms. On the other hand, because of the simplicity of the search mechanism of the greedy algorithm, the benefit of the specific search mechanism of the dispatching optimization algorithm developed in this chapter is more intuitive.

The framework of the implemented greedy algorithm is briefly depicted in Figure 5-1. More detailed procedures of the greedy algorithm are attached in Appendix I. As stated above the basic timetable serves as the initial solution of the dispatching optimization algorithm, in which inherent conflicts between trains exist. A conflict is embodied in a knock-on delay of a train at a certain block section on the microscopic and mesoscopic level or at a certain infrastructure node on the macroscopic level. A conflict can be resolved with a series of dispatching actions (see Section 5.2.2), and a candidate solution will be accordingly generated. The candidate solution is depicted by train priority sequences on all open track sections and loop tracks. Afterward, it will be simulated, and the train priority sequences on all open track sections and loop tracks are explicitly controlled according to the candidate solution (see Section 5.1). Based on the protocol of the simulation, the value of a certain dispatching objective function can be calculated. If the solution quality is improved, replace the initial solution with the current candidate solution; otherwise the next candidate solution will be calculated. The sum of knock-on delays is used as the dispatching objective function here for example. This process will be executed iteratively until the terminate specification is fulfilled or the solution quality cannot be further improved with any of the candidate solutions. In some cases, it is also possible that the initial solution is already better than any of the candidate solutions, which implies that First Come First Serve (FCFS) is the best dispatching principle at this moment. In such cases the optimization process will terminate directly. The maximum number of iterations and computation time are implemented as two terminate specifications, and they can be set by the users before the optimization process starts. In reality, the total computation time available for dispatching tasks should be dynamically set depending on online factors, such as the time span of prediction, the number of trains involved in conflicts, the scope of delay propagation and the urgency of solving conflicts.

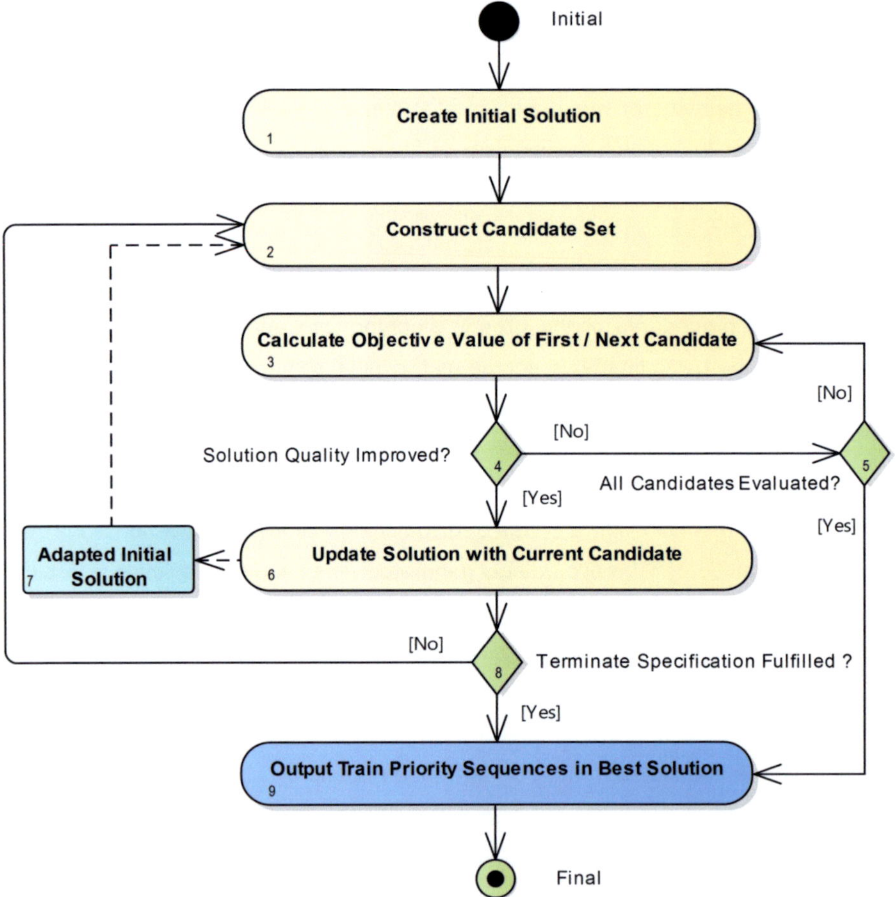

Figure 5-1: The Framework of Greedy Algorithm

In principle, the dispatching optimization algorithm follows the same procedure no matter on which description level (microscopic or mesoscopic or macroscopic level). Compared to that on the microscopic and mesoscopic level, some small details of the optimization algorithm may have to be accordingly adjusted on the macroscopic level. For instance, the location of knock-on delay can be accurate to block sections on the microscopic and mesoscopic level, but only to infrastructure nodes (i.e. loop track, junction node, open track section) on the macroscopic level. Moreover, the node attribute – loop non track does not exist on the macroscopic level due to the high level of abstraction, so the part of the dispatching optimization algorithm concerning loop non track should excluded on the macroscopic level. The dispatching optimization algorithm on the macroscopic level can be treated as a simplified version of that on the microscopic and macroscopic level. In order to avoid content duplication, only the dispatching optimization algorithm on the microscopic and mesoscopic levels will be explained in detailed in the following context.

5.1 Train Priority Sequence Control

In order to correctly reproduce a candidate solution, the train priority sequences on infrastructure nodes have to be explicitly controlled during the simulation process. However, only FCFS dispatching principle is implemented in the simulation model developed in Chapter 3, which cannot meet the requirement of dispatching optimization tasks. So the simulation model is modified and new constraints on train priority sequence are integrated. In the simulation model described in Chapter 3, the Banker's algorithm is implemented to regulate the sequence of trains running through the infrastructure resources in order to avoid deadlock problem. The Banker's algorithm has the authority to change the path of a train and the train priority sequences on infrastructure resources. Therefore, it may have contradiction with the new sequence control logic. To avoid sequence control contradiction, the Banker's algorithm is removed, and the train priority sequences on infrastructure resources are completely controlled according to the given candidate solution. This requires that the candidate solution should be deadlock-free. Before a candidate solution is simulated, the train priority sequences defined by the solution will be analyzed. If circle wait situations happened, the candidate solution will be directly abandoned, and then continue to the next candidate solution.

The function of train priority sequence control is realized by using two auxiliary variables – the arrival and departure lists on a loop track or an open track section[22]. For a loop track or an open track section, the arrival list defines the sequence of trains entering it, and the departure list defines the sequence of trains departing from it. To integrate the function of train priority sequence control in the simulation model, the three procedures included in the workflow - request resources, allocate resources and proceed with simulation tasks – should be accordingly modified.

In the procedure of requesting resources, only if a resource requirement meets the constraints of the relevant arrival and departure lists, the requirement can be permitted; otherwise the requirement should be rejected. To find the relevant arrival and departure lists, two pieces of information should be prepared: the last physically occupied block section by the requester (named Block 1) and the required block section beyond Block 1 along the path of the requester (named Block 2). The node attribute of Block 1 is called current node, and the node attribute of Block 2 is called requested node in the following context. All possible node attributes of the current and requested nodes are enumerated as follows.

- open track section (virtual block section not included)
- open track section (virtual block section included)
- loop track
- loop non track
- junction

If the node attribute of the current node is any of the first three types, the resource requirement should fulfill the constraints of the arrival and departure lists on the very next loop track or open track section except two exceptions. The first exception is that the train will not leave the current node in the current time interval (e.g. Train 2 in Figure 5-2). In this case the node attributes of the current and the requested nodes are the same. The second exception is that the very next loop track or open track section is an open track section consisted of a virtual block section (e.g. Train 1 in

---

[22] The open track section exclusively consisted of free resources has nothing to do with train priority sequence control, since any train is allowed to enter such an infrastructure node at any time by definition.

Figure 5-2). In both case the resource requirement can be directly permitted, because there is no possibility of violation of the constraints on train priority sequence.

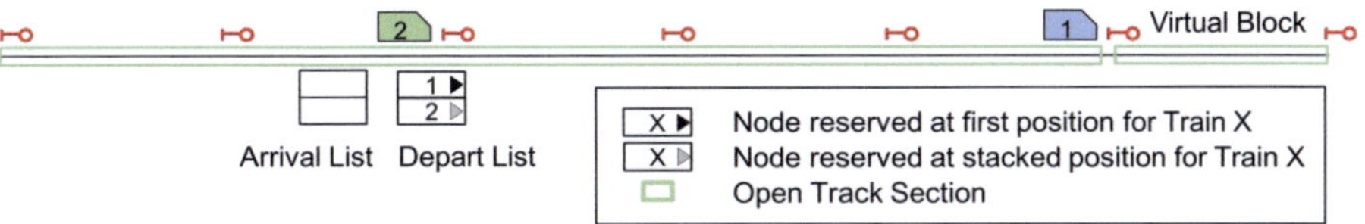

Figure 5-2: An Example of the Two Exceptions in Priority Sequence Control

Except the two situations explained above, the new resource requirement should be checked with the following method.

Step 1: Determine the very next node (i.e. loop track or open track section) based on the current node.

Step 2: If the very next node is not reserved at the first position for this train in the arrival list, go to Step 6. Otherwise go to the next step.

Step 3: If the very next node is reserved at the first position for this train in the departure list, go to Step 5. Otherwise go to the next step.

Step4: If the running direction of the train, which departs previously than this train in the departure list of the very next node, is opposite to the running direction of this train, go to Step 6. Otherwise go to Step 5.

Step 5: Permit the new resource requirement.

Step 6: Reject the new resource requirement

If the node attribute of the current node is any of the last two types in the list above, it implies that the priority sequence constraints on the very next open track section were definitely previously fulfilled. So the resource requirement should be permitted.

In the procedure of resource allocation, the permitted resource requirements only need to pass the conflict-free test; because the candidate solution must be deadlock-free. In the procedure of proceeding with simulation tasks, after the position of a train is updated, the relevant arrival and departure list may also have to be updated. Updating of arrival list is triggered by new allocated resource. The node attributes of the new allocated resources[23] should be collected. If any of these nodes is a loop track or open track section and the first position of its arrival list is reserved for the train, the train will be removed from the arrival list. Updating of departure list is triggered by released resources. The node attributes of the released resources should also be collected. If any of them is a loop track or open track section and the rear of the train has completely left the loop track or open track section, the train will be removed from its departure list.

With the modification elaborated above, now the simulation model can accept external dispatched timetables (i.e. candidate solutions) and accordingly run simulations.

5.2   Candidate Set

To construct the candidate set, every conflict present in the initial solution will be attempted to be resolved. Once a candidate solution is generated, it has to be simulated in order to evaluate its solution quality. Simulation is a very time-consuming task. It is necessary to rank the conflicts, so that important conflicts can be resolved primarily. The evaluation method of the priority of conflict will be explained Section 5.2.1, and conflicts resolution will be discussed in Section 5.2.2.

5.2.1   Conflict Evaluation

The dispatching optimization algorithm aims to minimize the summation of knock-on delays of trains in a given timetable. So, as the concrete manifestation of conflicts, the individual knock-on delays should be ranked according to their contribution to the final results. Two criteria are used to evaluate the contribution of a knock-on delay: the knock-on delay itself and the influence of the knock-on delay on further conflicts. The first criterion represents the direct contribution of the knock-on delay to the final result, and the second criterion represents the indirect contribution of the knock-on delay to the final result.

For the assessment of delay propagation scope a simplified method has been developed in Section 4.1. The basic logic of delay propagation has been revealed: the

---

[23] The node attributes of the new allocated overlaps are not included. Overlaps can be released in case of scheduled and unscheduled stops. The repeated changes of overlap occupations may violate train priority sequence control, so overlaps should be excluded for train priority sequence control.

knock-on delay of a train leads to the delay of the train itself, and the delay of the train can lead to the knock-on delay of another train. In the following context of this section, the relationship between the source knock-on delay and the delay of a train will be referred to as intra-train relationship, and the relationship between the delay of a train and the resulting knock-on delay of the other trains as inter-train relationship. To quantify the influence of a knock-on delay on further conflicts, just knowing the trains involved in conflicts is not enough. So the method is modified and expanded to describe the behavior of delay propagation more accurately. In the following context, the new method will be explained based a small example.

For intra-train relationship, take the small case shown in Figure 5-3 as an example. Due to the hindrance of the train Z1 on the block section B1, the train Z2 obtained a knock-delay $tw_{2,0}$ of two minutes on B0. As the direct result of $tw_{2,0}$, the delays of Z2 on the further block sections $td_{2,i}$ were increased to two minutes. Because $td_{2,i}$ is exclusively caused by $tw_{2,0}$, $tw_{2,0}$ takes 100% responsibility for $td_{2,i}$.

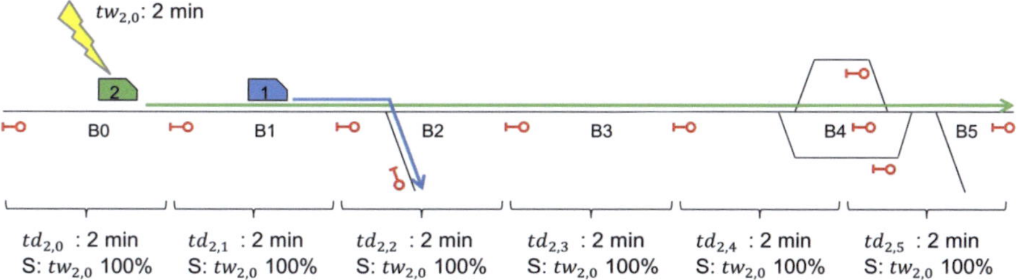

Figure 5-3: Relationship between Delay and Knock-on Delay of a Train (I)

When a delay is caused by several knock-on delays, the responsibility should be distributed to each concerned knock-on delay. It is supposed that Z2 obtained another knock-on delay $tw_{2,3}$ of three minutes on B3 because of the hindrance of Z3 on B4 (Figure 5-4). As a result, the delays of Z2 on B3, B4 and B5 were increased to 5 minutes. Now both $tw_{2,0}$ and $tw_{2,3}$ are the sources of $td_{2,3}$, $td_{2,4}$ and $td_{2,5}$, so responsibility should be proportionally distributed to each of them. Delay $tw_{2,0}$ takes 40% responsibility, and delay $tw_{2,3}$ takes 60% responsibility.

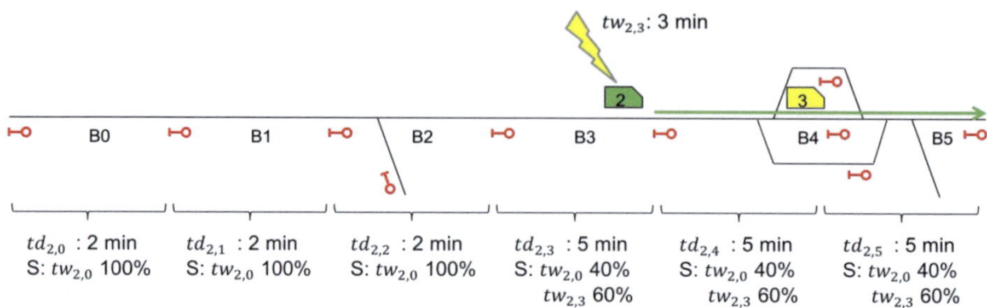

Figure 5-4: Relationship between Delay and Knock-on Delay of a Train (II)

Through responsibility distribution the relationship between delays and knock-on delays of a train can be accurately described, and the procedure of responsibility distribution is summarized below.

Step 1: calculate the knock-on delays of a train. It is supposed that there are n knock-on delays for this train. $(n \geq 1)$

Step 2: select the ith knock-on delay, enumerate the block sections from the one on which the knock-on delay occurred to the last second block section along the path of the train. $(i \in [1, n])$

Step 3: for each enumerated block section, a delay instance will be generated on it if the delay instance does not exist. A delay instance is identified by two attributes: the delayed train and the block section.

Step 4: record the ith knock-on delay in the source knock-on delay list of each enumerated block section.

Step 5: if all knock-on delays of the train have been analyzed, go to Step6, otherwise go to Step 2.

Step 6: check each block section along the path of the train. If a delay instance is defined on a block section, the responsibility of each source knock-on delay will be calculated with formula (5-1).

$$P^{Rb}_{td_{j,k}, tw_{j,i}} = \frac{tw_{j,i}}{\sum_{b=1}^{b=n_j^{block}} tw_{j,b} \cdot \chi_{\{l| tw_{j,l} \in S^{source}_{td_{j,k}}\}}} \tag{5-1}$$

Notation used:

$P^{Rb}_{td_{j,k},tw_{j,i}}$: Percentage of the responsibility of source knock-on delay $tw_{j,i}$ for delay

$td_{j,k}$

$tw_{j,i}$: Knock-on delay of train j on block section i

$td_{j,k}$: Delay of train j on block section k

$X_{\{l|l\in S_{td_{j,k}}\}}$: Indicator function, if the knock-on delay of train j on block section l $tw_{j,l}$

is the source of the delay of train j on block section $td_{j,k}$, it is equal to 1;

otherwise it is equal to 0.

n_j^{block}: Number of block sections along the path of train j

$S^{source}_{td_{j,k}}$: Set of source knock-on delays of delay $td_{j,k}$

According to the basic logic of delay propagation (inter-train relationship), the delay of a train may result in the knock-on delay of the other trains. As shown in Figure 5-5, three trains Z2, Z4 and Z5 are performing scheduled stops in station. When the scheduled dwell times are fulfilled, Z2, Z5 and Z4 will attempt to depart from the station in sequence. Due to the 5 minute delay of Z2 on B6, Z4 is hindered by Z2 for 4 minutes. Afterward Z2 is hindered by Z5 for another 1 minute. Thus, the knock-on delay of Z4 on B7 $tw_{4,7}$ is 5 minutes, and $td_{2,6}$ takes 80% (4 min / 5 min) and $td_{5,6}$ takes 20% (1min / 5 min) responsibility for it.

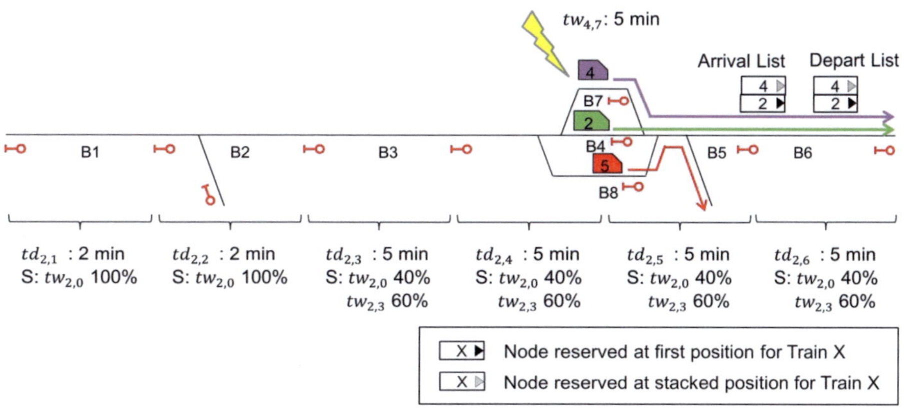

Figure 5-5: The Delay of a Train Leads to a Knock-on Delay of another Train

In principle the percentage of responsibility of a delay for the resulting knock-on delay can be determined with the following formula.

$$P^{Rb}_{tw_{j,i}, td_{l,k}} = \frac{t_{h_{j,i,l,k}}}{t_{h_{j,i}}}$$

(5-2)

Notation used:

$P^{Rb}_{tw_{j,i}, td_{l,k}}$: Percentage of responsibility of delay $td_{l,k}$ for knock-on delay $tw_{j,i}$

$t_{h_{j,i,l,k}}$:   Time period during which train j on block i had been hindered by train l on block k

$t_{h_{j,i}}$:   Time period during which train j on block i had been hindered by the other trains

In the simulation process, when a hindrance occurs, the most important relevant information will be recorded. It includes the IDs and positions of the hindered and hindering trains and the duration of the hindrance. This information will be outputted as a protocol of conflict relationship at the end of a simulation. Based on this protocol the relationship between delays and their resulting knock-on delays can be quickly reconstructed. In the reconstruction process a special case should be noted: a train may be hindered by another punctual train. In other words, the source delay of the knock-on delay may do not exist. In such a case, a fake delay should be created. A fake delay is a temporary variable, and introduced only to help calculating the percentage of responsibility of a delay for its resulting knock-on delay. In the calculation process, it is treated the same as a real delay.

Based on the reconstructed intra-train and inter-train relationships, the delay propagation diagram of a simulated timetable can be drawn accurately as shown in Figure 5-6. Along the propagation path of the knock-on delay $tw_{2,0}$ (marked with bold black line), the influence of $tw_{2,0}$ on further conflicts can be calculated as follows:

$$Inf_{tw_{2,0}} = P^{Rb}_{td_{2,6}, tw_{2,0}} \cdot P^{Rb}_{tw_{4,7}, td_{2,6}} \cdot tw_{4,7}$$
$$+ P^{Rb}_{td_{2,6}, tw_{2,0}} \cdot P^{Rb}_{tw_{4,7}, td_{2,6}} \cdot Inf_{tw_{4,7}}$$

(5-3)

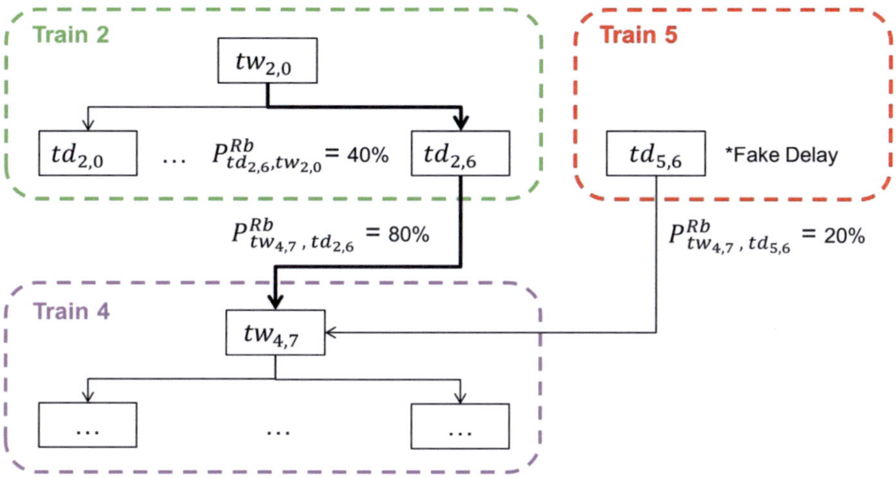

Figure 5-6: Delay Propagation Diagram

The calculation of the influence of a knock-on delay on further conflicts is recursive process, for instance $Inf_{tw_{4,7}}$ should also be determined with the same method and substituted in Formula 5-3. The recursion ends when all propagation paths terminate.

The knock-on delay and influence of the knock-on delay on further conflicts should be normalized at first, and then combined into a comprehensive priority indicator for the importance of knock-on delays as shown in Formula 5-4. Furthermore, the relative importance of the knock-on delay and the influence on further conflicts may vary in different system states, so a variable C_{tw} is introduced to represent their relative importance.

$$Pri_{tw_{j,i}} = C_{tw} \cdot tw_{j,i}' + (1 - C_{tw}) \cdot Inf_{tw_{j,i}}' \qquad (5\text{-}4)$$

Notation used:

$Pri_{tw_{j,i}}$:   Priority of knock-on delay $tw_{j,i}$

C_{tw}:   Relative importance of weighted knock-on delay ($C_{tw} \in [0,1]$)

$tw_{j,i}'$:   Normalized weighted knock-on delay of train j on block section i

$Inf_{tw_{j,i}}'$:   Normalized influence of a knock-on delay $tw_{j,i}$ on further conflicts

For a given system state, C_{tw} should be according adjusted to achieve the best performance of the dispatching optimization algorithm. System state classification will be elaborated in Chapter 6, and an application of the state-dependent dispatching algorithm will be represented in Chapter 7.

5.2.2　Conflict Resolution

For a selected knock-on delay, the suitable dispatching actions are chosen based on its circumstances. Three dispatching actions including passing, overtaking and re-platforming are implemented for a conflict resolution. As stated in Section 5.1, only the train priority sequences on loop tracks and open track sections are controlled, so only these variables are considered for the dispatching task.

Both overtaking and passing are used to exchange the priority sequences between a train and its immediately previous train on a loop track or an open track section. Overtaking applies to the situation of two trains with a successive movement, while passing applies to the situation of two trains with an opposite movement. Both of them are realized through the "swap" move operation. By a swap move operation, the priority indexes of two selected trains on a loop track or an open track section will be exchanged, and the priority indexes of the other trains on this loop track or open track section remain the same. Compared to overtaking or passing, swap does not require that the two selected trains are directly next to each other in the arrival and departure list. An example of the swap operation is shown in Figure 5-7: on a loop track, the priority indexes of Train 1 and Train 2 are exchanged through a swap operation. Because overtaking and passing use the same movement operation logic, they will be together referred to as "overtaking" in the following context.

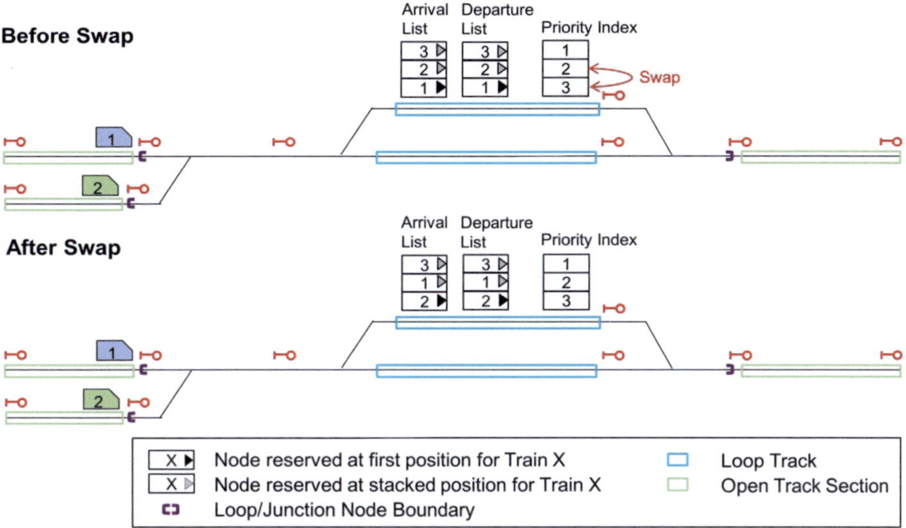

Figure 5-7: An Example of a Swap Operation on a Loop Track

By replatforming the position of a train will be changed from its original loop track to an alternative loop track. Replatforming is realized through the "insert" move operation. An insert operation consists of three steps as shown in Figure 5-8. Firstly, the train will be removed from the arrival and departure list on the original loop track (e.g. Train 2). Secondly, the arrival and departure lists on both the original and alternative loop tracks should be accordingly updated if necessary. The insert position on the alternative loop track should be pre-given. At last, the train will be inserted into the arrival and departure list on the alternative loop track.

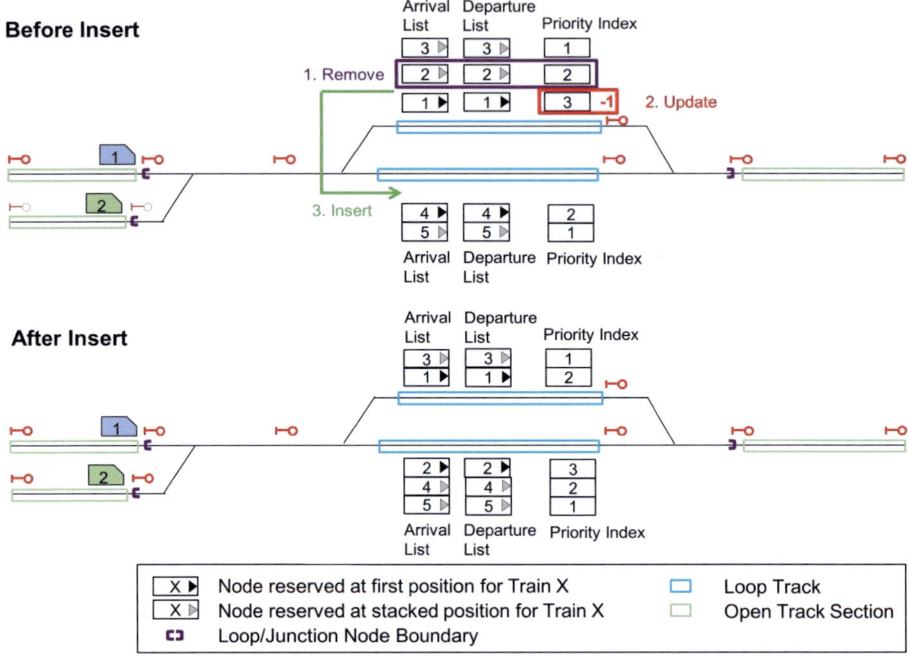

Figure 5-8: An Example of an Insert Operation

For the correct choose of dispatching actions, the circumstance of the selected knock-on delay is analyzed from two perspectives – the infrastructure node where the knock-on delay occurred (i.e. the location of the knock-on delay) and the next dispatching relevant infrastructure node. As described in Section 5.2.1 a knock-on delay has two importance attributes – the delayed train and the block section where the knock-on delay occurred. Based on the block section and the path of the train, these two infrastructure nodes concerned can be determined. For different locations of the knock-on delay, the corresponding suitable dispatching actions may be different. There are four types of infrastructure node, so the circumstances are accordingly divided into four categories at first:

–   Category I: the knock-on delay locates on a loop track

–   Category II: the knock-on delay locates on a loop non track

–   Category III: the knock-on delay locates on a junction

–   Category IV: the knock-on delay locates on an open track section

The type of the next dispatching relevant infrastructure node also has influence on the correct choose of dispatching actions, so each category will be further divided into several subcategories. Each subcategory has a specific suitable dispatching action. The details of these four categories will be discussed separately in the following context.

For category I, the type of the next dispatching relevant infrastructure node should be determined primarily as shown in Figure 5-9. When it is an open track section, the delayed train will be arranged to overtake its immediately previous train on this open track section. This case is defined as subcategory I-C. If the overtaking action is successfully performed, the new relative priority sequences of these two trains will dominate their relative priority sequences on the other part of their common macro path. In other words, their relative priority sequences on the other part should be checked and adjusted in order to avoid the violation of priority sequence consistency. If the overtaking action is not successfully performed (e.g. no previous train exists), this knock-on delay will be skipped and the next knock-on delay will be selected according to the ranking of all knock-on delays.

When the next dispatching relevant infrastructure node is a loop track in another loop (not the loop where the knock-on delay locates) and the delayed train attempts to enter the loop (the next block section of the train enters the loop), the replatforming action will be performed through the insert operation at first. Only if the execution of the replatforming action failed, the overtaking action will be carried out. This case is defined as subcategory I-A. To execute the insert operation, the optimal alternative loop track and the insert position on it should be determined. For an alternative loop track, the expected occupation time of the delayed train on it should be estimated. By counting the overlapping between the expected occupation time of the delayed train the actual occupation time of pre-existing trains on the alternative loop track, the corresponding expected conflict can be calculated. The alternative loop track with the minimum conflict is the optimal one. The insert position on the optimal alternative track can be determined by sorting the start occupation times of all involved trains. With the optimal alternative track and the insert position on it the insert operation can be executed.

Unlike the overtaking action, even though the replatforming action is successfully performed, the relative priority sequences between the delayed train and pre-existing trains on the other path of their common macro path will not be accordingly adjusted. On the one hand, theoretically, the relative priority sequences between the delayed train and every pre-existing train on the alternative loop track should be checked and adjusted. However, excessive modification of the initial solution probably deteriorates the solution quality or even generates infeasible solutions. On the other hand, adjusting the relative priority sequences between trains can become very time-consuming by taking all pre-existing trains and the delayed train into consideration. In summary, adjusting relative priority sequences between trains on their common macro will not be implemented in this optimization model.

If the next dispatching relevant infrastructure node is a loop track but the delayed train does not attempt to enter it (the next block section of the train does not enter the loop), the overtaking action will be executed. This case is defined as subcategory I-B. The replatforming action could also be implemented for this subcategory. However, as stated above the expected conflict on all alternative loop tracks should be estimated in order to determine the optimal alternative loop track and the correct insert position on it. The reliability of the estimation increases with decreasing distance between the delayed train and the loop. So a conservative approach is adopted herein. Only if the delayed train attempts to enter the loop, execution of the replatforming action is allowed.

The classification of subcategories and designation of suitable dispatching action for Category I is summarized in Figure 5-9. This category is very representative, and it will be used as an important component part of the other three categories. With consideration of the special constraints of the other three categories, additional subcategories may have to be introduced.

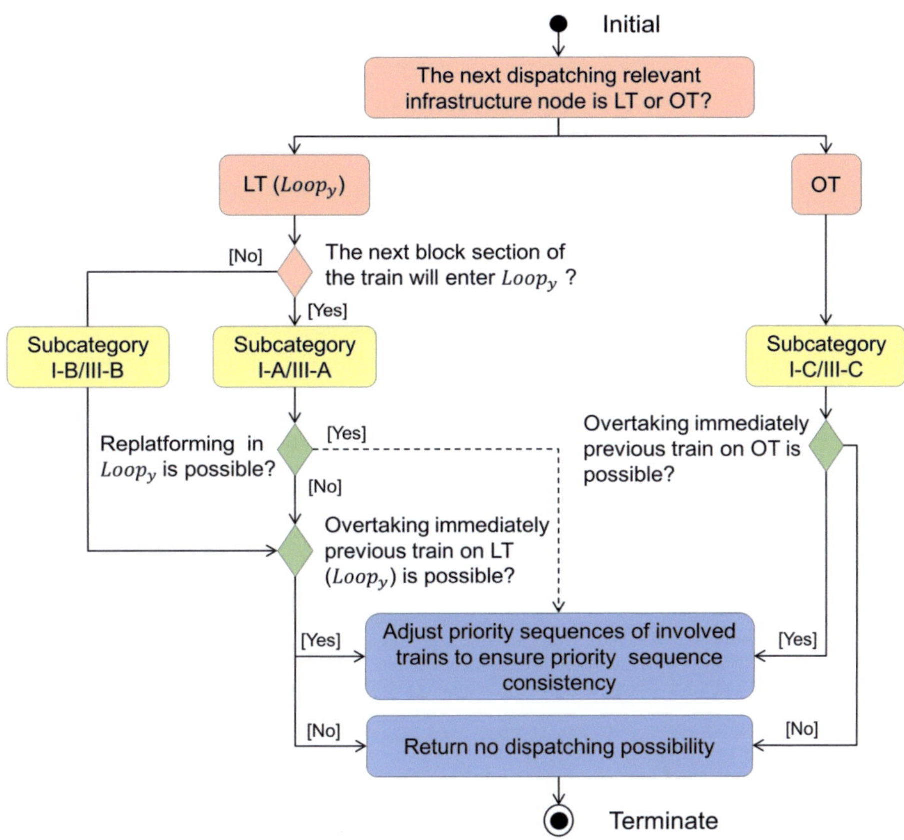

Figure 5-9: Classification of Subcategories for Category I of Conflicts Circumstance

For Category II, the knock-on delay is located on the loop non track in a loop. The delayed train is either approaching or leaving a certain loop track in the loop. These two cases should be treated in different ways. Under conditions of approaching, the delayed train will be arranged to overtake its immediately previous train on the loop track. This subcategory is defined as subcategory II-D. Similar to the subcategory I-B, considering the reliability of the estimation of conflicts on alternative loop tracks, it is not recommended to use the replatforming action for this subcategory. When the knock-on delay occurred, the train has already entered the loop. The occupation times of the other concerned trains in this loop are the results of the interactions between the delayed train and the other trains. The interactions between trains will change as the scheduled loop track of the delayed train changes. As a result, the occupation times of trains in the loop probably changes as well. Therefore, the replatforming action is not implemented for subcategory II-D. Under conditions of leaving, the type of next dispatching relevant infrastructure node should be identified primarily. Moreover, if the next dispatching relevant infrastructure node is a loop, whether the delayed train attempts to enter the loop should also be determined. This case can be treated exactly in the same manner as Category I (see Figure 5-9).

No special constraint exists for Category III, so it is treated in the same manner as Category I. For Category IV, because the knock-on delay is located on an open track section, it is necessary to determine whether the hindrance is located on the current open track section or on the next infrastructure node. The former case is defined as subcategory IV-D, and in this case the delayed train will be arranged to overtake its immediately previous train on this open track section. In the latter case, the further classification of subcategories and designation of suitable dispatching action follows the same procedure as that in Category I (see Figure 5-9).

6 Classification of System States

Dispatching decisions are highly related to the system state of the whole investigation area. Even though the same objective function is used, the most suitable dispatching algorithm may differ in different system states. A multi-state concept is proposed based on the hypothesis that each state has its specific most suitable dispatching algorithm and a schematic representation of it is shown in Figure 6-1. The system state is characterized by three attributes: the traffic flow / density of traffic (the X-axis), the average waiting time per train (the Z-axis) and the proportion of the delayed trains (the Y-axis) over the whole investigation area. In the theory of capacity research, the waiting time function reveals that the mutual hindrance of trains (average waiting time) increases with the increase of traffic flow. The recommended area of traffic flow (the green line on the X-axis) represents the optimal traffic flow of the investigation area; below this level, the infrastructure is not efficiently used and above this level the traffic is congested [Hertel, 1992; Pachl, 2002; Martin, 2014]. In this way three different sections of traffic flows are separated. Furthermore, for each traffic flow (when the traffic flow is fixed), the average waiting time may be higher or lower than the statistical mean value of the approximated waiting time function (the red curve on the X-Z plane in Figure 6-1). The area below and above the waiting time function is separated. Likewise, for each traffic flow, the situations where a small proportion of trains are delayed with large knock-on delays and where a large proportion of trains are delayed with medium or small knock-on delays should be treated differently in the dispatching process. Thus the delayed train proportion is also introduced into the concept. The boundary value of the delayed train proportion between the two different situations is a function of the traffic flow. As an example, the function is illustrated as the blue line on the X-Y plane in Figure 6-1. With the 3 levels of traffic flow, 2 levels of average waiting time and 2 levels of delayed train proportion, the three dimension system is finally divided into 12 subspaces. The traffic flow used herein does not exceed the maximum throughput capacity[24]. The subspace with an ex-

---

[24] In [Chu, 2013], the maximum throughput capacity is defined as the average load in the stationary phase of the operation process, at which the maximum capacity research throughput is reached for a given infrastructure with a given coarse operating program while maintaining the train mix (i.e. structure of the operating program).

tremely high traffic flow (over the maximum throughput capacity) will be treated as a special state[25]. In this study, the different states are represented by the timetables with stochastic deviations generated with the existing software PULEIV [Martin et al. 2008a; Martin et al. 2008d, Martin et al. 2008c].

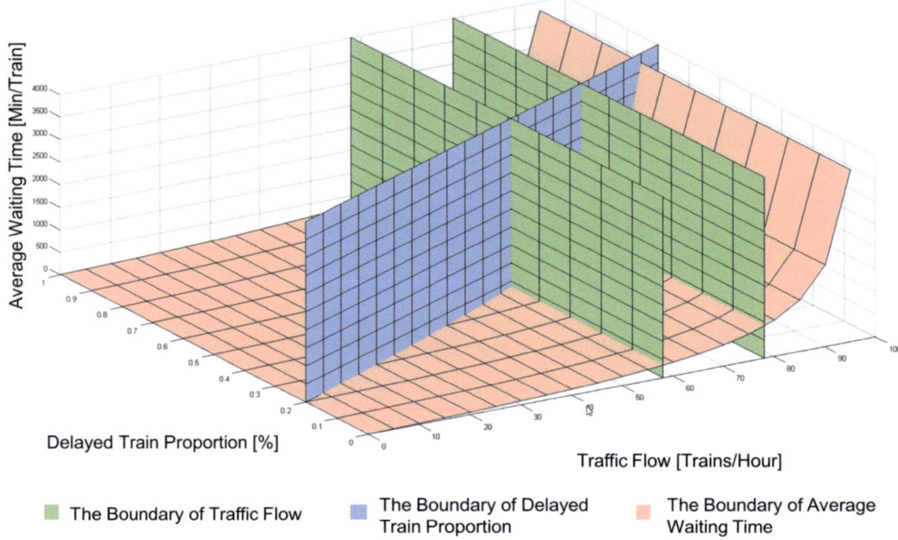

Figure 6-1: An Example of State Classification (based on: [Hertel, 1992]; [Martin and Chu, 2012])

Following the theory of capacity research, the maximum throughput capacity, waiting time function and recommended area of traffic flow can be determined by means of simulation tools and the software PULEIV. The detailed description of the theory of capacity research can be found in [Martin, 2014] and [Chu, 2014]. For a better understanding, a brief description is attached in Appendix III. In this chapter, only the new introduced boundary line (i.e. delayed train proportion) will be elaborated.

In this approach, the delayed train proportion is defined based on the theory of capacity research, so zero second[26] is taken as the threshold value to distinguish de-

---

[25] The detailed boundary conditions of the 13 system states are attached in Appendix II.

[26] As a preliminary study, this research takes zero second as the criterion to judge the delayed train. In real railway operation, delays of trains on-site counted cannot achieve such accuracy; even trains with minor delays can also be regarded as punctual. Therefore, in further researches, it is recommended to use a larger value to judge delayed trains.

layed trains. If the waiting time of a train is more than zero second, it is counted as a delayed train in this approach. For a timetable f the number of delayed trains N_{d_f} can be counted:

$$N_{d_f} = \sum_{z=1}^{z=E_f} \chi_{\{l|t_{w_{l,f}}>0\}}(z)$$

(6-1)

Notation used:

N_{d_f}:   Number of delayed train in timetable f

E_f: Entry load of timetable f

$\chi_{\{l|t_{w_{l,f}}>0\}}(z)$:   Indicator function, if the waiting time of a train z is greater than 0,

it is equal to 1, otherwise equal to 0.

In the theory of capacity research, trains belonging to entry load are considered in the calculation of waiting time function, to keep the consistency of methodology delayed train proportion is also counted based on trains belonging to entry load. Entry load is defined as the number of trains which run into the investigation area or depart at their original stations located inside the investigation area during the specified time period for analysis according to the pre-given schedule. Delayed train proportion for a timetable f is calculated as follows.

$$P_{d_f} = \frac{N_{d_f}}{E_f}$$

(6-2)

Notation used:

P_{d_f}:   Delayed train proportion of timetable f

After the delayed train proportion of each timetable was calculated, the delayed train proportions of the timetables that have the same entry load are averaged to reduce the deviation of data points.

$$P_{d_E} = \frac{\sum_{f=1}^{n_{ges}} P_{d_f} \cdot \chi_{\{l|E_l=E\}}(f)}{\sum_{f=1}^{n_{ges}} \chi_{\{l|E_l=E\}}(f)} \tag{6-3}$$

Notation used:

P_{d_E}: Delayed train proportion of a certain entry load E

n_{ges}: Total number of timetables

$\chi_{\{l|E_l=E\}}(f)$: Indicator function, if the entry load of a timetable f is equal to a given entry load E, it is equal to 1, otherwise equal to 0.

As a result, one entry load and its average delayed train proportion correspond to a data point along the trend curve of average delayed train proportion. Through several experiments with different operating programs on a reference infrastructure network, it was found that average delayed train proportion is linearly correlated with entry load. The result of an experiment is shown in Figure 6-2. Linear regression is used to model the trend line of average delayed train proportion in this example. The coefficient of determination R^2 is high (over 95%), which indicates average delayed train proportion and entry load have a strongly positive linear relationship.

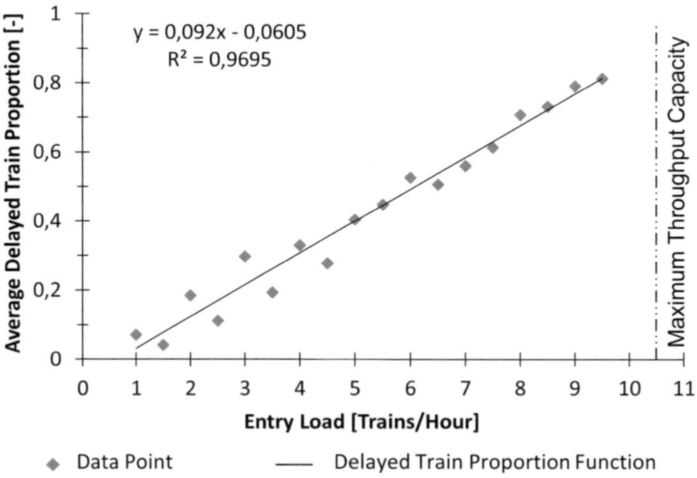

Figure 6-2: An Example of the Average Delayed Train Proportion Function

So the model function of average delayed train proportion can be summarized as follows.

$$P_d = \begin{cases} 0, & c_1 \cdot E + c_2 < 0 \\ c_1 \cdot E + c_2, & 0 \leq c_1 \cdot E + c_2 \leq 1 \\ 1, & c_1 \cdot E + c_2 > 1 \end{cases} \tag{6-4}$$

Notation used:

P_d: Average delayed train proportion

E: Entry load ($E \in (0,\ \text{maximum throughput capacity})$)

c_1, c_2:   Parameters of delayed train proportion function

7 Evaluation of the Influences of Dispatching on the Capacity and Operation Quality

The capacity and operation quality are always depending on each other and their relationship is represented by the results of capacity research, which include the maximum capacity, the waiting time function and the recommended area of traffic flow (abbr. OLB). In order to quantitatively evaluate the influences of dispatching on the results of capacity research, several rounds of capacity analysis with the implementation of different dispatching algorithms were carried out on two reference examples. Three dispatching algorithms were considered: First Come First Serve, the state-dependent dispatching algorithm, and a state-independent dispatching algorithm.

For the state-dependent dispatching algorithm, the state-dependent variable (i.e. the relative importance of knock-on delay) will be adjusted in each state to achieve the best performance of the dispatching algorithm. State-independent dispatching algorithm is also based on the state-dispatching algorithm introduced in Chapter 5, but the relative importance of knock-on delay and influence on further conflicts is determined to optimize the overall performance of the dispatching algorithm in all the timetables without considering the state classification. State-independent dispatching algorithm can also be called one state dispatching algorithm by definition.

7.1 Investigation Scenarios

To execute capacity research with implementation of different dispatching algorithms, the software PULEIV was used to generate a large amount of timetables with stochastic deviations based on a basic operating program and a reference infrastructure network. The considered time interval of each generated timetable is 6 hours, which is composed of three parts (2 hours each): preheating time, investigated time period and cool down time [Chu, 2014]. The investigated time period represents the rush hour in the investigated area. The macroscopic view of the infrastructure network of the reference example is shown in Figure 7-1.

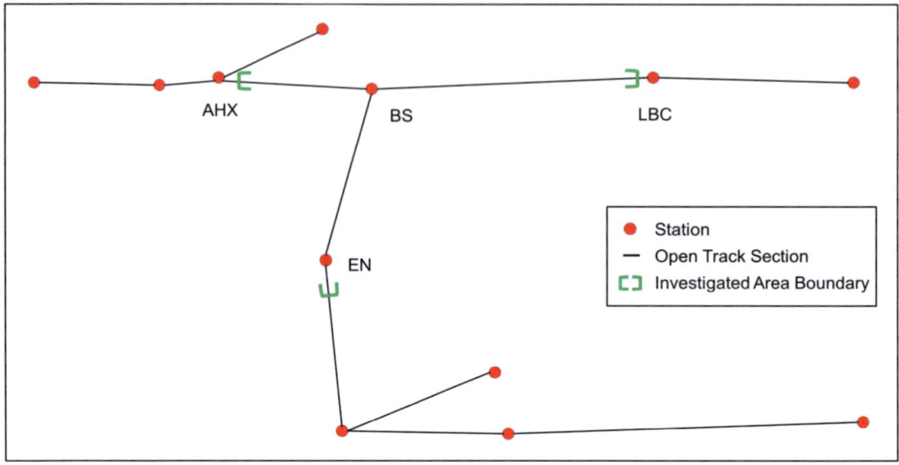

Figure 7-1: The Sketch of the Complete Infrastructure Network of the Reference Example

A large amount of previous simulation experiments have been carried out, and the experiment results showed that conflicts between trains (especially opposing and merging conflicts) mainly occurred in the area between the stations AHX, BS, EN and LBC. So, this area was used as the investigated area in this study. In order to ensure the accuracy of the results of capacity analysis, the investigated area was zoomed in, and simulated with the microscopic model. The microscopic view of the investigated area is presented in Figure 7-2.

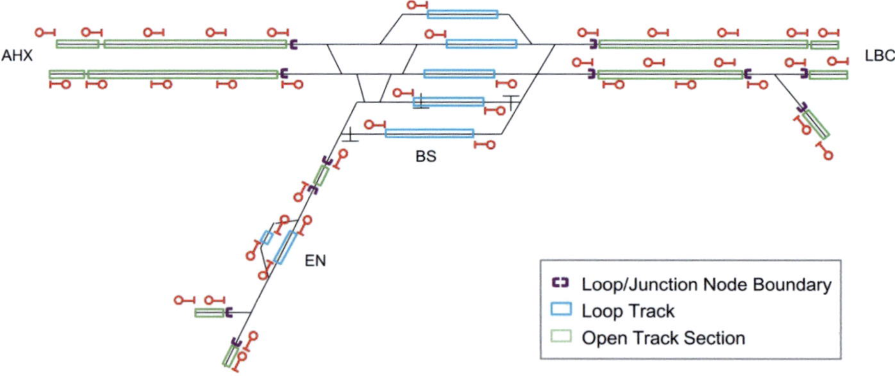

Figure 7-2: Microscopic View of the Investigated Area

Two typical conflicts situations – opposing and merging conflicts on common infrastructure resources – were used as two investigation scenarios to evaluate the influence of dispatching on the results of capacity research. The basic operating program depicting the conflict situation of opposing and the one depicting the conflict situation of merging are shown in Figure 7-3. For the conflict situation of opposing, two groups of trains (short-distance passenger train and freight train) were defined for each running direction (from Station EN to LBC or from Station LBC to EN). There is a high probability of occurring opposing conflicts on the infrastructure resources between Station BS and EN. For the conflict situation of merging, three groups of trains (short-distance passenger train, long-distance passenger train and freight train) were defined, all of which run towards the station LBC.

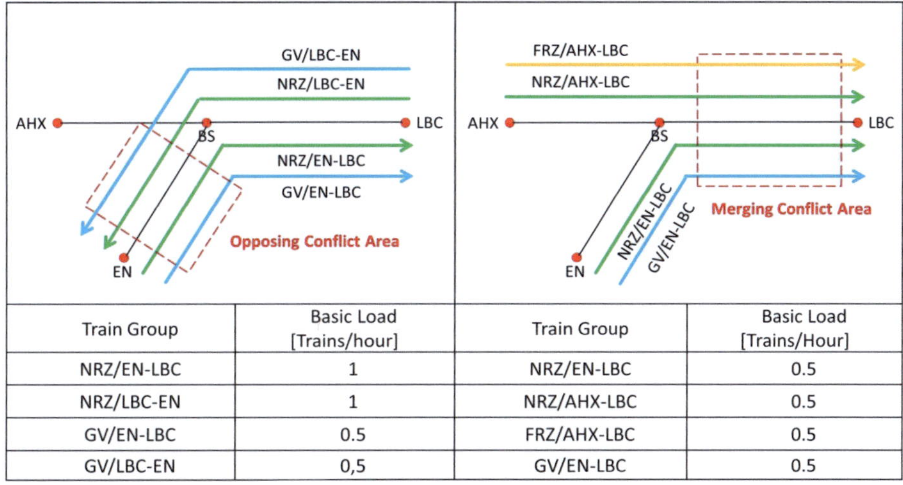

Train Group	Basic Load [Trains/hour]	Train Group	Basic Load [Trains/Hour]
NRZ/EN-LBC	1	NRZ/EN-LBC	0.5
NRZ/LBC-EN	1	NRZ/AHX-LBC	0.5
GV/EN-LBC	0.5	FRZ/AHX-LBC	0.5
GV/LBC-EN	0,5	GV/EN-LBC	0.5

Figure 7-3: Operating Program of Opposing and Merging Conflict Scenarios (FRZ: long-distance passenger train, NRZ: short-distance passenger train, GV: freight train)

7.2 Comparison of the Results of Capacity Research under Consideration of Different Dispatching Algorithms

For the investigation scenario of opposing conflicts, 512 timetables with the entry loads ranging from 0.3 Trains/Hour to 14.4 Trains/Hour are generated based on the basic operating program. The generated timetables were firstly simulated with implementation of the First Come First Serve dispatching algorithm. Based on the protocol of the simulations the results of capacity research including maximum throughput capacity, waiting time function and recommended area of traffic flow were calculated (marked in red in Figure 7-5). Additionally the curve of delayed train proportion was determined for the classification of system states.

To optimize the generated timetables with the state-dependent dispatching algorithm developed in Chapter 3, the relative importance of knock-on delay should be set in advance. The relative importance of knock-on delay, which is a continuous variable ranging from 0 to 1, was discretized into 11 quantiles using the following breakpoint 0, 0.1, $\cdots$, 0.9, 1. Each generated timetable was optimized 11 times, and each time the relative importance of knock-on delay in the state-dependent dispatching algorithm was set to a different value. In the optimization process, only the trains belonging to the entry load, whose scheduled departure times at their initial station located in the investigated area or at the boundary of the investigated area are within the investigated time period, are taken into account, because the optimization of train runs within the preheating time and cool down time will invalidate the system classification defined in Chapter 6. Dispatching actions could change the structure of the operating program within the preheating time and cool down time, which results in a different system state within the investigated time period. So, the timetable of 6 hours is simulated as a whole, and only the section within the investigated time period is optimized. The maximum computation time of the optimization algorithm was set to 15 minutes for all timetables. The maximum computation time of the optimization algorithm should be set depending on the computational capabilities, time constraints and other on-site factors. For on-line dispatching, the dispatching time horizon is limited, such as the next 30 minutes, because train runs are likely to be disturbed in a long operations planning horizon [Jacobs, 2010]. In most cases the identified conflicts should be resolved in a short time. However, timetables of 6 hours have to be simulated iteratively in the optimization model for capacity research. So a relative long maximum computation time was set. The optimization model runs on a PC equipped with a processor Intel Core i5-4670 (3.40GHz), 8G RAM and Windows 7 operating system. The performance of the dispatching optimization algorithm on 4 test cases with different traffic flows (6 Trains/Hour, 9 Trains/Hour, 12 Trains/Hour, 15 Trains/Hour) are presented in Figure 7-4, and the test cases used belong to the investigation scenario of opposing conflicts. In the following figures the symbol "P" refers to the relative importance of knock-on delay in the dispatching optimization algorithm. For instance, P0.5 represents that the relative importance of knock-on delay is 50%. It can be seen that:

- for timetables with a relative low traffic flow (e.g. 6 Trains/Hour in Figure 7-4), all possible settings of the relative importance of knock-on delay are capable of guiding the dispatching optimization algorithm to find the same local optimum;

- for timetables with an extremely high traffic flow (e.g. 15 Trains/Hour in Figure 7-4), even though the final solutions found by the dispatching optimization algorithm with the guidance of different settings of the relative importance of knock-on delay are different, the quality of the solutions does not differ significantly;

- only in case of medium and relative high traffic flows (e.g. 9 Trains/Hour and 12 Trains/Hour in Figure 7-4), the performance of the dispatching optimization algorithm differs significantly in consideration of different settings of the relative importance of knock-on delay.

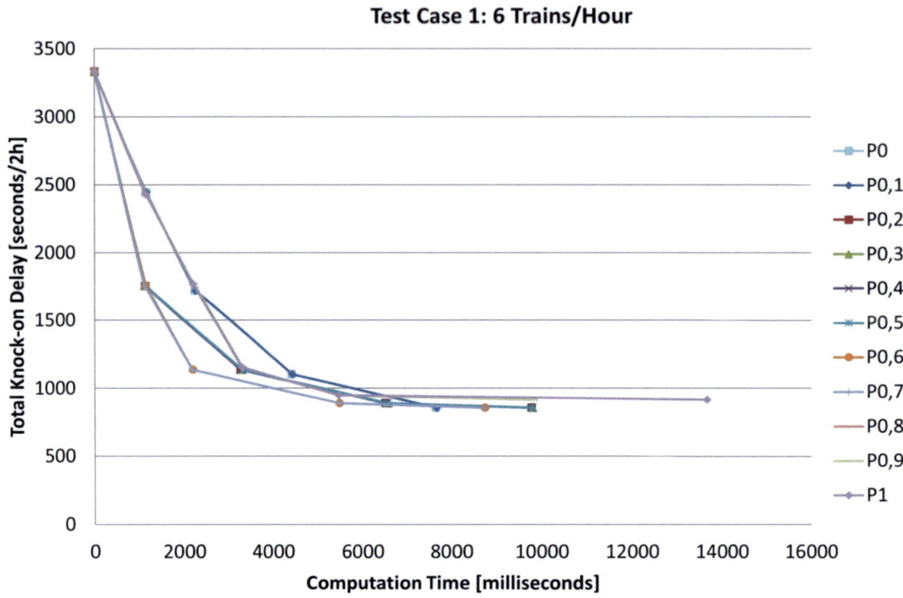

Test Case 1: 6 Trains/Hour
Total Knock-on Delay [seconds/2h]
Computation Time [milliseconds]
P0
P0,1
P0,2
P0,3
P0,4
P0,5
P0,6
P0,7
P0,8
P0,9
P1

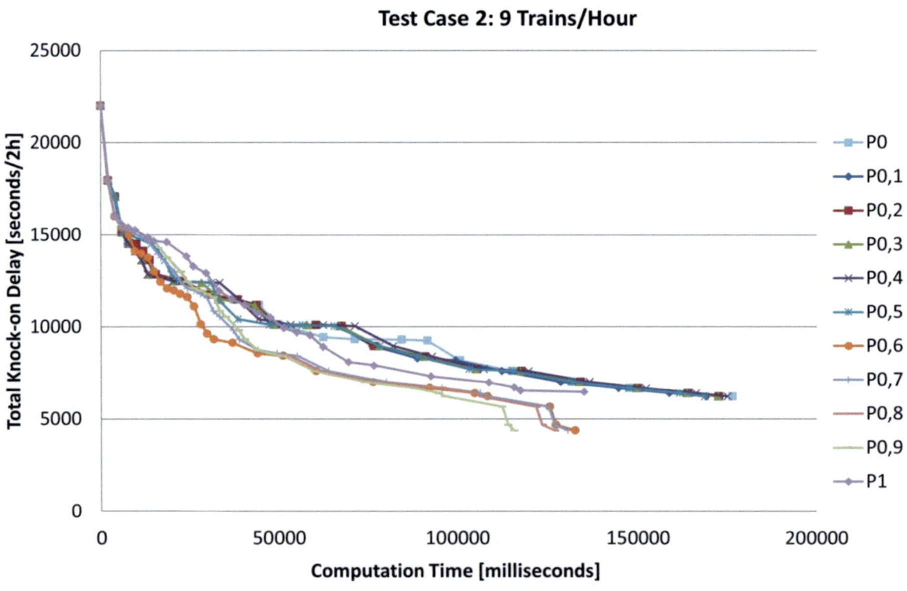

Test Case 2: 9 Trains/Hour
Total Knock-on Delay [seconds/2h]
Computation Time [milliseconds]
P0
P0,1
P0,2
P0,3
P0,4
P0,5
P0,6
P0,7
P0,8
P0,9
P1

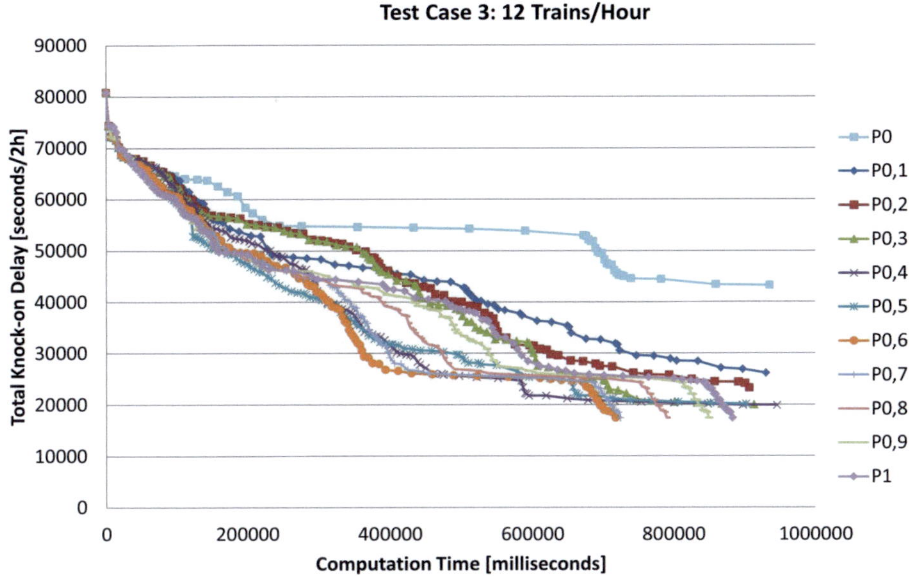

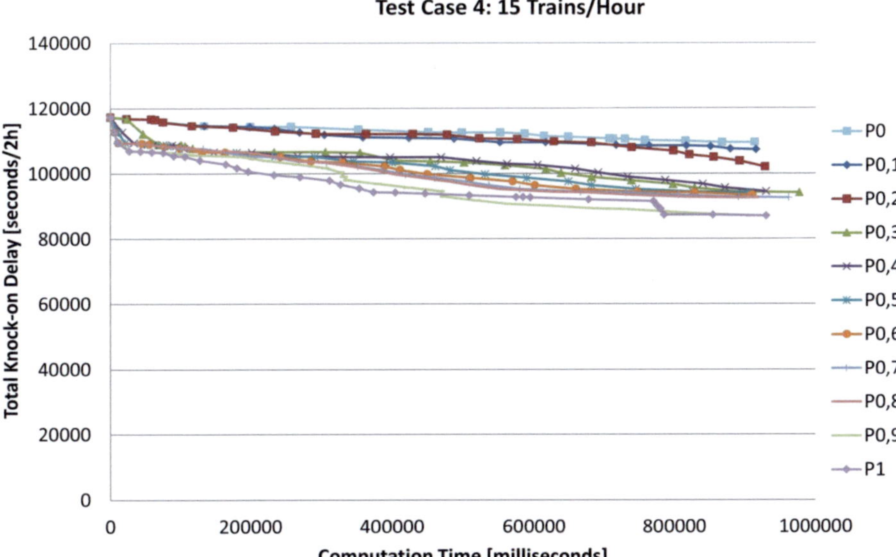

Figure 7-4: Performance of the Dispatching Optimization Algorithm – Opposing Conflicts (P: relative importance of knock-on delay, for the definition it is referred to page 70)

Once a generated timetable had been optimized by the dispatching algorithm with a certain setting of the relative importance of knock-on delay, the important data were correspondingly recorded in the simulation protocol, which include the value of the dispatching objective function, average waiting time, entry load and exit load in the investigated time period. With the protocols of all generated timetables, optimal settings of the dispatching algorithm in states can be determined. When several equivalently optimal settings of the dispatching algorithm exist in a state, one of them will be taken as the final optimal setting of the dispatching algorithm in that state. It is also possible that no timetable exist in a certain state. In this case all settings are taken as equivalent. The final optimal settings in all states on this investigation scenario are shown in the Table 7-1.

State	Optimal Setting	State	Optimal Setting
State 1-4	1.0	State9	0.5
State 5-7	0.6	State 10-12	0.6
State 8	0.8	State 13	0.9

Table 7-1: Optimal Settings of the Relative Importance of Knock-on Delay in the Dispatching Optimization Algorithm in All States

Based on the optimal settings in each state, the corresponding entry loads, exit loads and average waiting times in the investigated time period of the included timetables were outputted for conducting a new round of capacity analysis. The results of capacity research with the implementation of the state-dependent dispatching algorithm are illustrated in Figure 7-5 (marked in green)[27].

For the state-independent or one state dispatching algorithm, the optimal setting of the relative importance of knock-on delay is 0.7 on this investigation scenario. Following the same process as above, a new round of capacity analysis was conducted, and the results are illustrated in Figure 7-5 (marked in blue).

---

[27] The theory of capacity research is briefly introduced in Appendix III. The numerical results of the capacity analysis for both investigation scenarios (i.e. opposing and merging conflicts) are attached in Appendix IV.

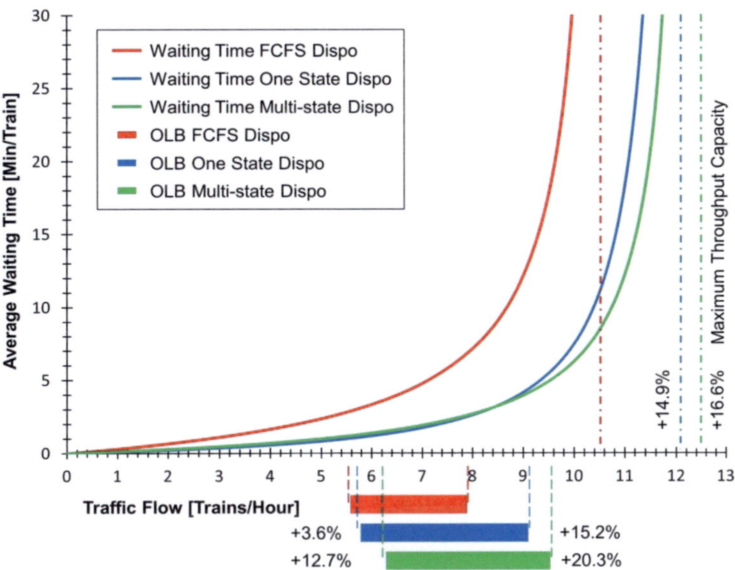

Figure 7-5: Results of Capacity Research for Different Dispatching Algorithms - Opposing Conflicts (OLB: Recommended Area of Traffic Flow)[28]

The experimental results shown in Figure 7-5 confirmed that with implementation of either the multi-state or one state dispatching optimization algorithm, the maximum throughput capacity and the recommended area of traffic are capable of being further improved comparing to the results of capacity research without dispatching optimization (i.e. FCFS dispatching principle). In this investigation scenario, by employing one state and multi-state dispatching optimization algorithm increases in maximum throughput capacity of 14.9% and 16.6% are achieved respectively. The comparison showed that the dispatching optimization algorithm is able to influence the maximum throughput capacity significantly in a suitable regulating area. However in some special cases, such as in undersized regulating areas where dispatching actions are hardly performable and their effect cannot be deployed [Martin et al., 2015], the effectiveness of dispatching algorithm could be limited. Compared to the one state dispatching optimization algorithm, the multi-state dispatching optimization algorithm further increased the maximum throughput capacity only by 1.7%. The timetables in

---

[28] The numerical results of the capacity analysis for the opposing conflicts scenario are attached in Appendix IV.

the system state 13 (extremely high entry load) is decisive for the determination of the new maximum throughput capacities with implementation of the dispatching optimization algorithm, and its performance varies slightly with different settings as illustrated in the test case 4 in Figure 7-4. Thus, there is no significant difference between the results. The same phenomenon also can be seen in the investigation scenario of merging conflicts as illustrated in Figure 7-6 (blue and green dash lines). In conclusion, both the one state and multi-state dispatching optimization algorithms can increase the maximum throughput capacity significantly on suitable regulating areas compared to the result for FCFS dispatching algorithm. Moreover, the maximum throughput capacity for multi-state dispatching algorithm could be greater than that for one state dispatching algorithm, but the difference between the two results is statistically not significant.

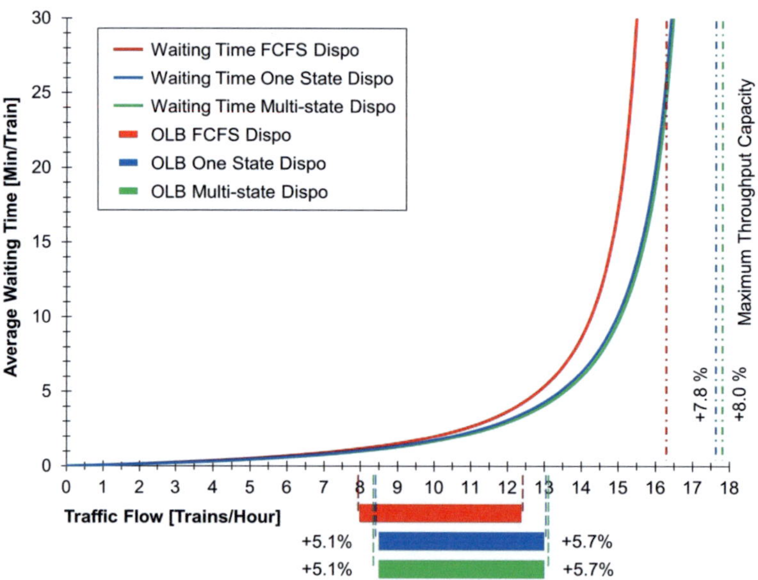

Figure 7-6: Results of Capacity Research for Different Dispatching Algorithms - Merging Conflicts[29]

In the investigation scenario of opposing conflicts, no matter which dispatching optimization algorithm is implemented, the average waiting times represented by the

---

[29] The numerical results of the capacity analysis for the merging conflicts scenario are attached in Appendix IV.

curve of waiting time function are substantially reduced (the blue and green curve in Figure 7-5), and the extent of reduction increases as the growth of the traffic flow. Accordingly, in both cases the recommended area of traffic flow shift in the direction of throughput capacity (the blue and green boxes in Figure 7-5), and the variation of the upper limit of the recommended area of traffic flow (increased by 15.2% and 20.3% respectively) is far larger than the variation of the lower limit (increased by 3.6% and 12.7% respectively). Taking the advantage of the proposed multi-state concept, it is capable of designating an optimal setting of the relative importance of knock-on delay for the dispatching optimization algorithm in each state (Table 7-1). On the contrary, the optimal setting of one state dispatching algorithm is a good compromise among different states (the optimal setting is 0.4 in this investigation scenario). Thus, the average waiting time for multi-state dispatching optimization algorithm, especially in case of medium and high traffic flow, are smaller than that for the one state dispatching optimization algorithm. Consequently, both the lower and upper limit of recommended area of traffic for the multi-state dispatching algorithm is about 6-9% greater than for the one state dispatching algorithm (from 3.6% to 12.7% and from 15.2% to 20.3% respectively). A better operational quality was achieved with the support of system state classification. In order to show the comparison of the experimental results more clearly, the results of both investigation scenarios are summarized in the following table.

	FCFS	One State	RC	Multi-state	RC
		Opposing Conflicts			
	FCFS	One State	RC	Multi-state	RC
Max [Trains/Hour]	10.51	12.08	+14.9%	12.25	+16.6%
L_{OLB} [Trains/Hour]	5.5	5.7	+3.6%	6.2	+12.7%
U_{OLB} [Trains/Hour]	7.9	9.1	+15.2%	9.5	+20.3%
		Merging Conflicts			
	FCFS	One State	RC	Multi-state	RC
Max [Trains/Hour]	16.35	17.66	+8.0%	17.62	+7.8%
L_{OLB} [Trains/Hour]	7.9	8.3	+5.1%	8.3	+5.1%
U_{OLB} [Trains/Hour]	12.3	13	+5.7%	13	+5.7%

* Max: maximum throughput capacity, L_{OLB}: lower bound of recommended area of traffic flow, U_{OLB}: upper bound of recommended area of traffic flow, RC: relative change

Table 7-2: Experimental Results for the two Investigation Scenarios

From the theoretical viewpoint, one state is only a special case of the multi-state concept, and therefore multi-state dispatching optimization algorithm is able to achieve at least the same performance as one state dispatching optimization algorithm, as can be seen at the investigation scenario of merging conflicts (see Figure 7-6). In this investigation scenario, a train is highly probably hindered when it attempts to enter the station BS (see Figure 7-2), consequently the delay will propagate to the following trains from the same initial station (EN or AHX). In addition, the number of trains from station EN and AHX are kept the same in the operating program. So, in most cases the high ranking knock-on delays have also high-ranking influences on further conflicts. Accordingly, the relative importance of knock-on delay became insensible to the changes of system state. Even in such extreme application conditions, the multi-state dispatching algorithm has the same performance as one state dispatching algorithm (the blue and green curve in Figure 7-6). Nevertheless, taking into account the convenience and ease of application, one state dispatching algorithm is recommended to be used in case of insensible dispatching algorithms.

8 Summary

Through the systematic analysis of the influence of the dispatching optimization algorithm on the results of capacity research, the hypothesis of system state classification postulated in Chapter 6 and the state-dependent dispatching algorithm proposed in Chapter 5 are model and theoretically confirmed to a large extent. Both the one state and multi-state dispatching optimization algorithm have a significant influence on the results of capacity research, which are embodied in the improvements of maximum throughput capacity and recommended area of traffic flow. Furthermore, one state dispatching optimization algorithm is only suitable in case that dispatching algorithms are insensible or system states are highly homogenized. In other cases, the multi-state dispatching optimization algorithm is recommended, with which a further improvement in the results of capacity research (especially the recommended area of traffic flow) can be obtained.

In practical applications, the optimal classification of system states and the optimal settings of the dispatching algorithm in the states should be case-specifically determined primarily, and then depending on the sensibility of the settings one state or multi-state dispatching algorithm could be implemented correspondingly. Moreover, this process for system state classification and optimization of dispatching algorithm settings in states is a general methodology also applicable for the other dispatching algorithms with state-dependent variables.

In general, the application of a state-dependent dispatching optimization algorithm can be summarized as illustrated in Figure 8-1. As an example, a state-dependent dispatching algorithm is formulated as $f = \alpha \cdot x_1 + \beta \cdot x_2 + \cdots + \gamma \cdot x_n$, in which α, β, $\cdots, \gamma$ are state-dependent variables. The optimal values of the state-dependent variables in different system states should be determined in advance through the systematic evaluation of the influences of dispatching on the results of capacity research (see Chapter 7). For on-line dispatching tasks or dispatching module integrated in simulation software, once potential conflicts are detected, the matching system state should be identified primarily. The dispatching algorithm is thereby configured with the optimal settings (i.e. $\alpha^*, \beta^*, \cdots, \gamma^*$). Accordingly, proper dispatching actions could be determined based on the pre-defined dispatching algorithm.

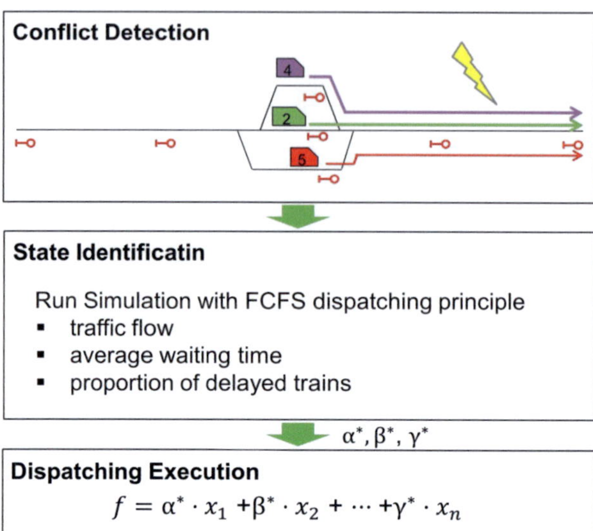

$$f = \alpha^* \cdot x_1 + \beta^* \cdot x_2 + \cdots + \gamma^* \cdot x_n$$

Figure 8-1: Application of a State-dependent Dispatching Optimization Algorithm

The influence of dispatching on the results of capacity research also indicates that it is necessary to combine timetable and operation simulation in extensive capacity research (see [Martin, 2014]), in order to reach a higher accuracy of capacity research. Last but not least, the findings of this study showed that capacity research is an appropriate standard to validate the effectiveness of dispatching algorithms. Even though different dispatching algorithms may have different dispatching objective functions, the theory of capacity research provides several unified and comprehensive evaluation criteria (i.e. the results of capacity research), which makes it possible to horizontally compare these algorithms. This could undoubtedly help simulation tools to improve their integrated dispatching algorithms.

Symbols

$AA_{Z_{Prev},Z_j,T}$	Arrive-arrive headway between the train Z_j and Z_{Prev} on open track section T	[Second]
$ACC_N^{micro \to meso}$	Aggregation accuracy of infrastructure node N from the microscopic level to the most detailed mesoscopic level	[Second]
ACC_{R_k,R_h}^{meso}	Aggregation accuracy of two occupation units R_k and R_h on the mesoscopic level	[Second]
C_{tw}	Relative importance of knock-on delay	[-]
c_1, c_2	Parameters of delayed train proportion function	[-]
$DA_{Z_{Prev},Z_j,T}$	Depart-arrive headway between the train Z_j and Z_{Prev} on open track section T	[Second]
$\Delta E(T^{OVLP,R_k+R_h})$	The relative change of expected total overlapping time period caused by combination of R_k and R_h	[Second]
$E(T_{t_2 \sim t_3,Z1,Z2}^{OVLP,R3})$	Expected total overlapping time period between Z1 and Z2	[Second]
E_f	Entry load of timetable f	[Trains/Hour]
$i[N]$	Index of the macroscopic node N in the macro-path of a corresponding train	[-]
$Inf_{tw_{j,i}}$	Influence of a knock-on delay $tw_{j,i}$ on further conflicts	[Second]
M	A number that is at least larger than the maximum of the aggregation accuracies	[Second]
n_j^{block}	Number of block sections along the path of train j	[-]
N_{d_f}	Number of delayed train in timetable f	[-]

$P^{R3}_{t_2 \sim t_3, Z1, Z2}$	Occurrence probability of the conflict situation between Z1 and Z2 on the occupation unit R3 in the time interval from t_2 to t_3	[-]
$P^{R3}_{t_2 \sim t_3, Zj}$	Blocking probability of the time interval $[t_2, t_3]$ for Zj on the occupation unit R3	[-]
$P^{Rb}_{td_{j,k}, tw_{j,i}}$	Percentage of the responsibility of source knock-on delay $tw_{j,i}$ for delay $td_{j,k}$	[-]
$P^{Rb}_{tw_{j,i}, td_{l,k}}$	Percentage of responsibility of delay $td_{l,k}$ for knock-on delay $tw_{j,i}$	[-]
$Pri_{tw_{j,i}}$	Priority of knock-on delay $tw_{j,i}$	[-]
P_{d_f}	Delayed train proportion of timetable f	[-]
P_{d_E}	Delayed train proportion of a certain entry load E	[-]
P_d	Average delayed train proportion	[-]
$S^{source}_{td_{j,k}}$	Set of source knock-on delays of delay $td_{j,k}$	[-]
$SV^{micro \rightarrow meso}_N$	Significance value of an infrastructure node N to be abstracted from the microscopic level to the most detailed mesoscopic level	[Second]
$SV^{meso}_{R_k, R_h}$	Significance value of a large mesoscopic occupation unit composed of R_k and R_h on the mesoscopic level	[Second]
T_{now}	Current execution time in the simulation model	[Second]
$TB_{i[N]-1, Z_j}$	Departing/passing time for train Z_j in the $(i[N]-1)^{th}$ node of its macro-path	[Second]
$TI_{i[N], Z_j}$	Real scheduled operational time for train Z_j in the $(i[N])^{th}$ node of its macro-path	[Second]
$TI^{ART}_{i[N], Z_j}$	Artificial scheduled operational time for train Z_j in the $(i[N])^{th}$ node of its macro-path	[Second]

$tw_{j,i}^{CG}, tw_{j,i}^{EG}$	Knock-on delay of train j on block section i in the controlled group (CG) or experimental group (EG)	[Second]
$t_{j,i}^{start,Ist}$	Actual start blocking time of train j on occupation unit i	[Second]
$t_{j,i}^{start,Soll}$	Scheduled start blocking time of train j on occupation unit i	[Second]
$t_{t_2 \sim t_3, Z1, Z2}^{OVLP,R3}$	Total overlapping time period between Z1 and Z2	[Second]
$tw_{j,i}$	Knock-on delay of train j on block section i	[Second]
$td_{j,k}$	Delay of train j on block section k	[Second]
$t_{h_{j,i,l,k}}$	Time period during which train j on block i had been hindered by train l on block k	[Second]
$t_{h_{j,i}}$	Time period during which train j on block i had been hindered by the other trains	[Second]
Z_j	Train Z_j	[-]
zj	Amount of block sections along the path of train j	[-]

Abbreviations

AA	Arrive-Arrive Headway
ATP	Automatic Train Protection
BS	Basic Structure
CG	Controlled Group
DA	Depart-Arrive Headway
EG	Experimental Group
FCFS	First Come First Serve
FIFO	First In First Out
FRZ	Long Distance Passenger Train
GV	Freight Train
LT	Loop Track
LNT	Loop non Track
NRZ	Short Distance Passenger Train
OLB	Recommended Area of Traffic Flow
OT	Open Track Section

Glossary

Basic structure

The maximum occupation unit allowed to be occupied by only one train simultaneously on a microscopic level without direction information. The boundaries of a basic structure can be the closest signal, signal releasing point, route releasing point as well as the borderline of the investigation area.

Initial delay

An initial delay is a type of primary delay that occurred at the boundary of the investigation areas.

Junction node

A junction node is consisted of at least one junction-type resource, but overtaking and passing are not able to be performed in it.

Knock-on delay

A delay caused by other trains due to either short headway times or late transfer connections [Pachl, 2002].

Loop node

A loop node is defined as an operational site where the dispatching actions - overtaking and passing – are able to be performed.

Loop non track

Except loop tracks the other part of a loop node refers to the loop non track.

Loop track

Loop tracks are the tracks on which scheduled or un-scheduled stops of trains can be performed inside loop nodes.

Maximum throughput capacity

The maximum throughput capacity is defined as the average load in the stationary phase of the operation process, at which the maximum capacity research throughput is reached for a given infrastructure with a given coarse operating program while maintaining the train mix [Chu, 2013].

Open track section	The infrastructures between nodes (i.e. loop node or junction node) are defined as open track sections
Original delay	An original delay is a type of primary delay that occurred at origin stations of affected trains which are located within the investigation areas.
Recovery time	A time supplement that is added to the pure running time to enable a train to make up small delays.
Unscheduled waiting time	Waiting time caused by hindrance during operational process, which is not included in the timetable.

Appendix I: Details of the Developed Greedy Algorithm

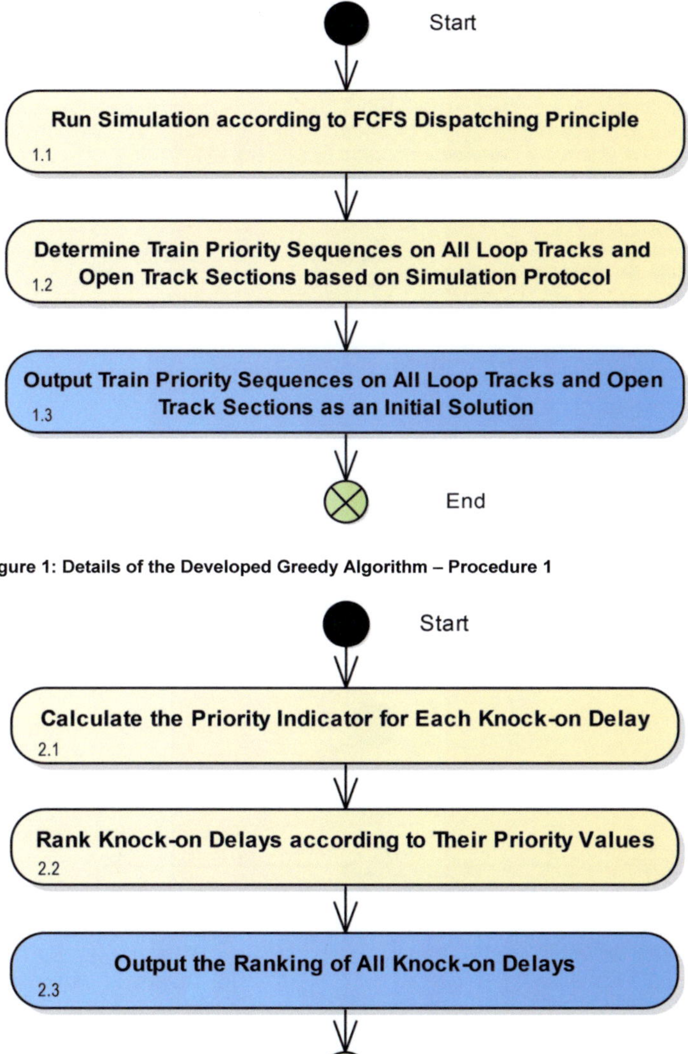

Appendix Figure 1: Details of the Developed Greedy Algorithm – Procedure 1

Appendix Figure 2: Details of the Developed Greedy Algorithm – Procedure 2

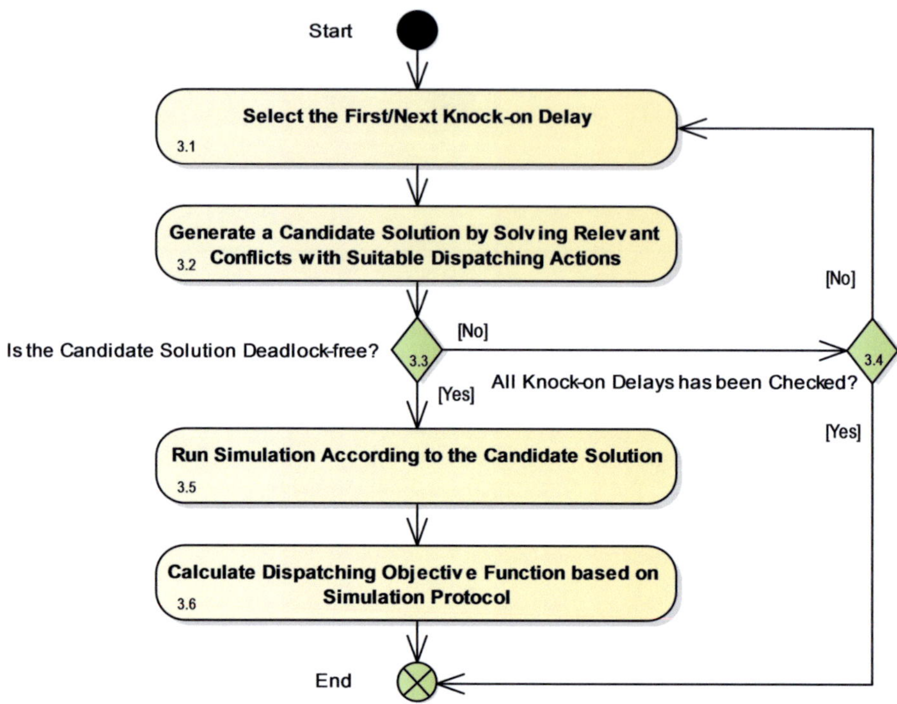

Appendix Figure 3: Details of the Developed Greedy Algorithm – Procedure 3

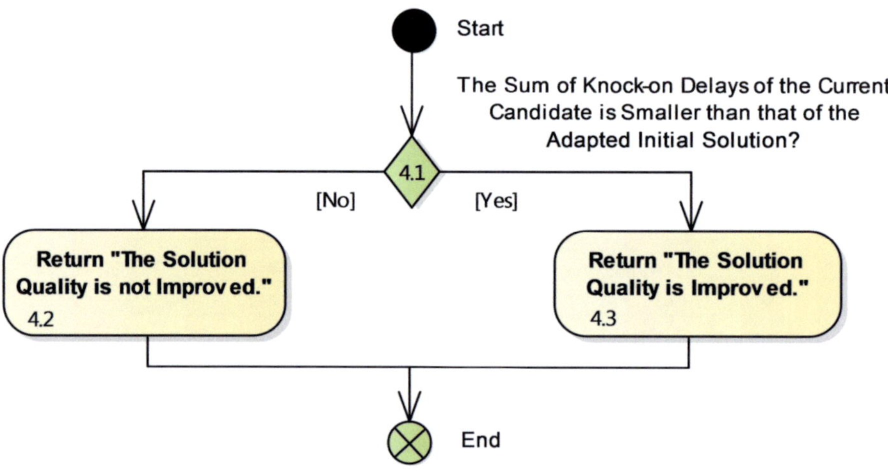

Appendix Figure 4: Details of the Developed Greedy Algorithm – Procedure 4

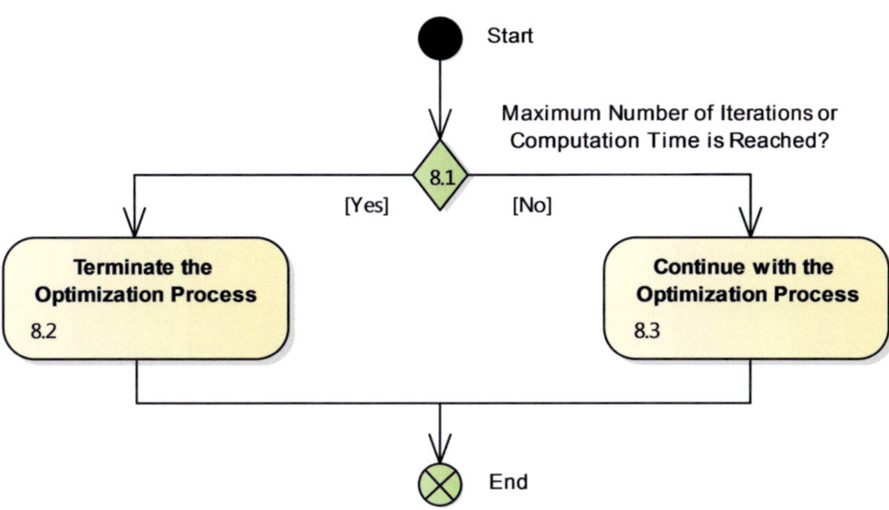

Appendix Figure 5: Details of the Developed Greedy Algorithm – Procedure 8

Appendix II: Settings of Boundaries for System States

Based on the theory of system state classification described in Section 6, traffic flow (or entry load E) is divided into four sections taking the lower limit (L_{OLB}) and upper limit (U_{OLB}) of recommended area of traffic flow (OLB) and maximum throughput capacity (Max) as the bounds; for a given timetable f with traffic flow E , the waiting time (tw_f) is judged by comparison with the statistical waiting time function (tw_E), and the delayed train proportion (P_{d_f}) is judged by comparison with the result of delayed train proportion function ($P_{d,E}$). In order to express more clearly, the 13 system states are named, and detailed boundary conditions are shown in Table A-1.

State	Traffic Flow	Delayed Train Proportion	Waiting Time
1	$E \leq L_{OLB}$	$P_{d_f} \leq P_{d,E}$	$tw_f \leq tw_E$
2	$E \leq L_{OLB}$	$P_{d_f} \leq P_{d,E}$	$tw_E < tw_f$
3	$E \leq L_{OLB}$	$P_{d,E} < P_{d_f}$	$tw_f \leq tw_E$
4	$E \leq L_{OLB}$	$P_{d,E} < P_{d_f}$	$tw_E < tw_f$
5	$L_{OLB} < E \leq U_{OLB}$	$P_{d_f} \leq P_{d,E}$	$tw_f \leq tw_E$
6	$L_{OLB} < E \leq U_{OLB}$	$P_{d_f} \leq P_{d,E}$	$tw_E < tw_f$
7	$L_{OLB} < E \leq U_{OLB}$	$P_{d,E} < P_{d_f}$	$tw_f \leq tw_E$
8	$L_{OLB} < E \leq U_{OLB}$	$P_{d,E} < P_{d_f}$	$tw_E < tw_f$
9	$U_{OLB} < E \leq Max$	$P_{d_f} \leq P_{d,E}$	$tw_f \leq tw_E$
10	$U_{OLB} < E \leq Max$	$P_{d_f} \leq P_{d,E}$	$tw_E < tw_f$
11	$U_{OLB} < E \leq Max$	$P_{d,E} < P_{d_f}$	$tw_f \leq tw_E$
12	$U_{OLB} < E \leq Max$	$P_{d,E} < P_{d_f}$	$tw_E < tw_f$
13	$Max < E$	-	-

Table A - 1: Boundary Conditions of System State

     The Influence of Dispatching on Capacity and Operation Quality of Railway Systems

Appendix III: Theory of Capacity Research

The detailed description of the theory of capacity research can be found in [Martin, 2014] and [Chu, 2014], and the theory is only introduced herein for a better understanding of the calculation method of maximum throughput capacity, waiting time function and recommended area of traffic flow. For the determination of these results by means of synchronous simulation several steps are necessary. The process is explained in the following context.

As the first step of capacity research the limits of the investigation area should be defined such that the sum of delays entering the operation process of the timetable (delays marked with red in Appendix Figure 6) should be approximately equal to the sum of delays exiting the operation process of the timetable (delays marked with orange in Appendix Figure 6). Delays entering the operation process of the timetable include initial delays of trains at the boundary of the investigation area and original delays of trains at their original stations located in the investigation area. Delays exiting the operation process include exit delays of trains at the opposite boundary of the investigation area and of trains at their destination stations located within the investigation area. To keep the sum of delays entering and exiting the operation process approximately equal, the recovery times of trains in the timetable should more or less compensate the original delays happening during the operation process (delays marked with gray in Appendix Figure 6).

Timetables with stochastic deviations are generated by using a stepwise increase of the train density while retaining the structure of a given base schedule. In timetables with stochastic deviations generated by the software PULEIV, the initial delays of trains at the boundary of the investigation area and the original delays of trains at the original stations located in the investigation area are modeled by negative exponential distribution. The other initial delays that occur during the operation process of the timetable in the investigation area, such as dwell time extension and running time extension are not considered because different recovery times are removed before the generation of timetables. These initial delays happening during the operation process should be compensated by different recovery times.

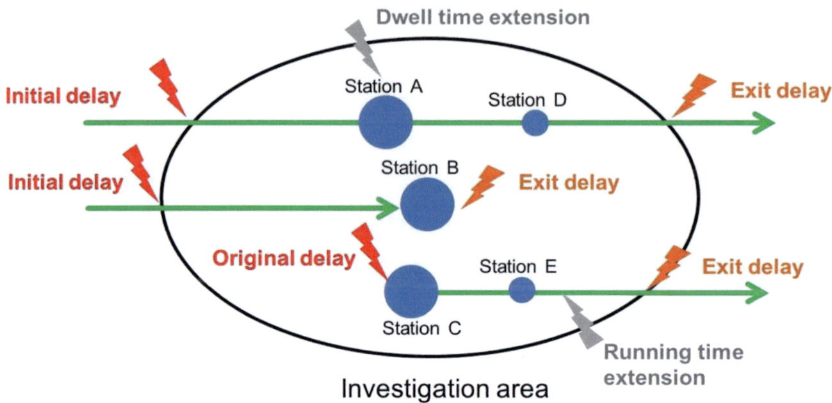

Appendix Figure 6: An Illustration of Different Types of Delays

During the simulation of each timetable, the start and end of occupation time on each basic structure along the path of each train are recorded, which are the basis for the calculation of maximum capacity, waiting time function and recommended area of traffic flow. The maximum capacity is determined by comparing the number of trains that enter (entry load) and leave (exit load) the investigation area during a specified time period for analysis. Entry load is defined as the number of trains which run into the investigation area or depart at their original stations located in the investigation area during the specified time period for analysis according to the pre-given schedule. Exit load is defined as the number of trains which run out of the investigation area or arrive at their destination stations located in the investigation area during the specified time period for analysis according to the actual operational situation. Entry load can be determined based on the pre-given schedule, and exit load can be counted as the number of trains whose end of occupation time on the last second[30] basic structure of their path are within the time period for analysis. If exit load of trains cannot further increase along with the increase of entry load of trains, maximum capacity is

---

[30] The last basic structure of a train path is a free resource which is extended from the investigation area for the convenience of modelling. The last second basic structure is the actual boundary of the investigation area.

reached[31]. The method for determination of maximum capacity is elaborated in [Chu, 2014].

Besides maximum capacity, average waiting time on each entry load is necessary for the calculation of waiting time function. The actual transport time of a train through the whole investigation area is the difference between the start of occupation time on the first basic structure and the end of occupation time on the last second basic structure of its path. The scheduled transport time of a train through the investigation area is calculated with the same method. The calculation methods of actual and scheduled transport times are described by:

$$t_{b_{z,f}}^{soll} = tB_{z,f,0}^{soll} - tE_{z,f,m}^{soll} \tag{A-1}$$

$$t_{b_{z,f}}^{ist} = tB_{z,f,0}^{ist} - tE_{z,f,m}^{ist} \tag{A-2}$$

Notation used:

$t_{b_i}^{soll/ist}$: Scheduled / actual transport time of a train z through the investigation area in timetable f

$tB_{i,0}^{soll/ist}$: Scheduled / actual start of occupation time on the first basic structure 0 of the path of train z in timetable f

$tE_{i,n}^{soll/ist}$: Scheduled / actual end of occupation time on the last second basic structure m of the path of train z in timetable f

The waiting time of a train $t_{w_{z,f}}$ is the difference between $t_{b_{z,f}}^{soll}$ and $t_{b_{z,f}}^{ist}$ as shown in following formula.

$$t_{w_{z,f}} = t_{b_{z,f}}^{ist} - t_{b_{z,f}}^{soll} \tag{A-3}$$

The average waiting time of a timetable is the average of the waiting times of the trains belonging to the entry load.

---

[31] As the inputs for determination of maximum capacity, entry loads and exit loads should be divided by the length of the time period for analysis, in order to ensure that the unit is trains per hour.

$$t_{w_f} = \frac{\sum_{z=1}^{z=E_f} t_{w_{z,f}}}{E_f} \qquad\qquad (A\text{-}4)$$

Notation used:

t_{w_f}: Average waiting time per train of timetable f

$t_{w_{z,f}}$: Waiting time of train z in timetable f

E_f: Entry load of timetable f

To reduce the deviation of data points from waiting time function, average waiting time for each entry load is calculated. The average waiting times of the timetables that have the same entry load are averaged.

$$t_{w_E} = \frac{\sum_{f=1}^{f=n_{ges}} t_{w_f} \cdot \chi_{\{l\,|E_l=E\}}(f)}{\sum_{f=1}^{f=n_{ges}} \chi_{\{l\,|E_l=E\}}(f)} \qquad\qquad (A\text{-}5)$$

Notation used:

t_{w_E}: Average waiting time of a certain entry load E

n_{ges}: Total number of timetables

$\chi_{\{l\,|E_l=E\}}(f)$: Indicator function, if the entry load of a timetable f is equal to a given entry load E, it is equal to 1, otherwise 0.

The average waiting time per train for an entry load corresponds with a data point along the curve of waiting time function. The average waiting times highly depend on the entry load. The higher the entry load, the higher the average waiting time is. Eventually the graph of waiting time function converges towards the maximum capacity as illustrated in Appendix Figure 7.

With maximum capacity and data points along the curve of waiting time function, the waiting time function can be determined according to the model function (A-6) [Martin, 2014; Martin and Chu, 2012]:

$$EW_t = t_w = \frac{a \cdot \eta}{(1 - \eta)^b} \qquad \text{(A-6)}$$

Notation used:

EW_t :　　　(Statistical) Expected waiting time function

t_w :　　　Average waiting time per train (because of standardization)

a, b :　　　Parameters of the waiting time function

η :　　　Utility factor of capacity

The utility factor of capacity results from:

$$\eta = \frac{\text{Traffic flow } n_{Zug} \text{ [number of trains/period of time]}}{\text{Maximum capacity LF [number of trains]}} \qquad \text{(A-7)}$$

For the adaption of the model function, the optimization toolbox of software MATLAB can be used, and the best-fit values of the parameters a, b will be automatically determined with the integrated curve fitting function [Chu, 2014].

The relative sensitivity of waiting time and its minimum value are calculated with the formulas (A-8) and (A-9) [Martin, 2014]. The minimum value is the lower limit of the recommended area of traffic flow for the investigation area under the given operation program.

$$EMPF_{rel} = \frac{t_w{}'}{t_w} \qquad \text{(A-8)}$$

$$EMPF_{rel}{}' \stackrel{!}{=} 0 \;\rightarrow\; \text{Minimum} \qquad \text{(A-9)}$$

The upper limit of the recommended area of traffic flow is the maximum value of traffic energy, and it is determined with the formula (A-10) and formula (A-11).

$$E_b = \frac{\eta}{1 + \dfrac{t_w}{t_b}} \qquad \text{(A-10)}$$

$$E_b{}' \stackrel{!}{=} 0 \;\rightarrow\; \text{Maximum} \qquad \text{(A-11)}$$

Notation used:

E_b :　Traffic energy

t_b :　Average transport time per train

The results of capacity research including maximum capacity, waiting time function and recommended area of traffic flow are illustrated in Appendix Figure 7.

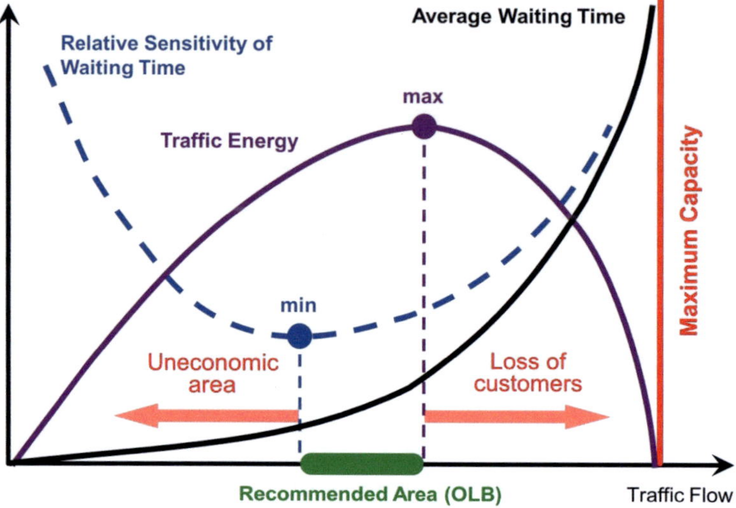

Appendix Figure 7: Recommended Area of Traffic Flow (OLB) [Hertel, 1992]

Appendix IV: Numerical Results of the Capacity Analysis

Investigation Scenario of Opposing Conflicts

First Come First Serve Dispatching Algorithm

Maximum Throughput Capacity: 10.51 [Trains/Hour]

Recommended Area of Traffic Flow: 5.5 – 7.9 [Trains/Hour]

Waiting Time Function:

$$ET_w = t_w = \frac{2.943 \cdot \dfrac{E}{10.51}}{(1 - \dfrac{E}{10.51})^{0.8111}}$$

Notation used:

EW_t : Statistical expected waiting time function [Min/Train]

t_w :   Average waiting time per train (because of standardization) [Min/Train]

E: Entry load [Trains/Hour]

One State Dispatching Algorithm

Maximum Throughput Capacity: 12.08 [Trains/Hour]

Recommended Area of Traffic Flow: 5.7 – 9.1 [Trains/Hour]

Waiting Time Function:

$$ET_w = t_w = \frac{1.036 \cdot \dfrac{E}{12.08}}{(1 - \dfrac{E}{12.08})^{1.231}}$$

Multi-state Dispatching Algorithm

Maximum Throughput Capacity: 12.25 [Trains/Hour]

Recommended Area of Traffic Flow: 6.2 – 9.5 [Trains/Hour]

Waiting Time Function:

$$ET_w = t_w = \frac{1.466 \cdot \dfrac{E}{12.25}}{(1 - \dfrac{E}{12.25})^{0.9779}}$$

Investigation Scenario of Merging Conflicts

First Come First Serve Dispatching Algorithm

Maximum Throughput Capacity: 16.35 [Trains/Hour]

Recommended Area of Traffic Flow: 7.9 – 12.3 [Trains/Hour]

Waiting Time Function:

$$ET_W = t_W = \frac{1.117 \cdot \dfrac{E}{16.35}}{(1 - \dfrac{E}{16.35})^{1.133}}$$

One State Dispatching Algorithm

Maximum Throughput Capacity: 17.66 [Trains/Hour]

Recommended Area of Traffic Flow: 8.3 – 13 [Trains/Hour]

Waiting Time Function:

$$ET_W = t_W = \frac{1.012 \cdot \dfrac{E}{17.66}}{(1 - \dfrac{E}{17.66})^{1.281}}$$

Multi-state Dispatching Algorithm

Maximum Throughput Capacity: 17.62 [Trains/Hour]

Recommended Area of Traffic Flow: 8.3 – 13 [Trains/Hour]

Waiting Time Function:

$$ET_W = t_W = \frac{1.017 \cdot \dfrac{E}{17.62}}{(1 - \dfrac{E}{17.62})^{1.268}}$$

Bibliography

[Alwadood et al, 2012]

Alwadood, Zuraida; Shuib, Adibah; Abd.Hamid, Norlida: *A Review on Quantitative Models in Railway Rescheduling.* In: International Journal of Science & Engineering Research, Vol. 3, Issue 6, 2012

[Brünger and Dahlhaus, 2014]

Brünger, Olaf; Dahlhaus, Elias: *Running Time Estimation.* In: Ingo Arne Hansen, Jörn Pachl (eds.): Railway Timetabling & Operations, Eurailpress, Hamburg, 2014

[Cacchiani et al, 2013]

Cacchiani, Valentina; Huisman, Dennis; Kidd, Martin; Kroon, Leo; Toth, Paolo; Veelenturf, Lucas; Wagenaar, Joris: *An Overview of Recovery Models and Algorithms for Real-time Railway Rescheduling.* In: Econometric Institute Report EI2013-29, 2013

[Chibana and Pierreval, 2010]

Mouelhi-Chibani Wiem; Pierreval, Henri: *Training a Network to Select Dispatching Rules in Real Time.* Computers & Industrial Engineering 58, 249-256, 2010

[Chu, 2014]

Chu, Zifu: *Modellierung der Wartezeitfunktion bei Leistungsuntersuchungen im Schienenverkehr unter Berücksichtigung der transienten Phase.* Dissertation, Universität Stuttgart, 2014

[Corman et al., 2016]

Corman, F.; D'Ariano, A.; D.Marra, A.; Pacciarelli, D.; Samà, M.: *Integrating Train Scheduling and Delay Management in Real-time Railway Traffic Control.* In: Transport Research Part E (article in press), 2016

[Corman and Meng, 2013]

Corman, Francesco; Meng, Lingyun: *A Review of Online Dynamic Models and Algorithms for Railway Traffic Control.* In: IEEE, 2013.

[Corman et al., 2011]

Corman, F.; D'Ariano, A.; Pacciarelli, D.; Pranzo, M.: *Dispatching and Coordination in Multi-area Railway Traffic Management.* RT-DIA-190, 2011

[Cui and Martin, 2011]

Cui, Yong; Martin, Ullrich: *Multi-scale Simulation in Railway Planning and Operation.* In: Promet-Traffic and Transportation, Vol.23, No.6, 2011, pp. 511-517

[Cui, 2010]

Cui, Yong: *Simulation-Based Hybrid Model for a Partially Automatic Dispatching of Railway Operation.* Dissertation, Universität Stuttgart, 2010

[D'Ariano and Pranzo, 2008]

D'Ariano, A.; Pranzo, M.: *An Advanced Real-time Train Dispatching System for Minimizing the Propagation of Delays in A Dispatching Area under Severe Disturbances.* In: Networks and Spatial Economics, Vol.9, Issue 1, 2008, pp. 63-84

[D'Ariano, 2008]

D'Ariano, A.: *Improving Real-Time Train Dispatching: Models, Algorithms and Applications.* PhD Thesis, TU Delft, 2008

[Fabris et al., 2014]

Fabris, Stefano; Longo, Giovanni; Medeossi, Giorgio; Pesenti, Raffaele: *Automatic Generation of Railway Timetables based on A Mesoscopic Infrastructure Model.* In: Journal of Railway Transport Planning & Management, Vol. 4, Issues 1-2, August-October 2014, pp. 2-13

[Gille et al., 2008]

Gille, A.; Klemenz, M.; Siefer, Th.: *Multiscaling Analysis.* In: WIT Transactions on The Built Environment, Vol 103, 2008

[Hertel, 1992]

Hertel, Günter: *Die maximale Verkehrsleistung und die minimale Fahrplanempfindlichkeit auf Eisenbahnstrecken.* In: Eisenbahntechnische Rundschau, Nr. 10, Jg. 41,1992, S. 665–671

[Horng, 2006]

Horng, Horng-Chyi: *Comparing Steady-state Performance of Dispatching Rule-pairs in Open shops*, International Journal of Applied Science and Engineering, 2006

[Kecman et al., 2012]

Kecman, Pavle; Corman, Francesco; D'Ariano, Andrea, Goverde, Rob M.P.: *Rescheduling Models for Network-wide Railway Traffic Management,* Research Project: Model-Predictive Railway Traffic Management (Project No. 11025), 2012

[Kettner et al., 2003]

Kettner, Michael; Sewcyk, Bernd; Eickmann, Carla: *Integrating Microscopic and Macroscopic Models for Railway Network Evaluation.* In: Proceedings of the European Transport Conference (ETC), Strasbourg, France, 2003

[Liang, 2017]

Liang, Jiajian: *Metaheuristic-based Dispatching Optimization Integrated in Multi-scale Simulation Model of Railway Operation.* Dissertation, Universität Stuttgart, 2017

[Marinov and Viegas, 2011]

Marinov, Marin; Viegas, Jose: *A Mesoscopic Simulation Modelling Methodology for Analyzing and Evaluating Freight Train Operations in Railway Network.* In: Simulation Modelling Practice and Theory, 2011

[Martin et al., 2015]

Martin, Ullrich; Liang, Jiajian; Cui, Yong: *Einfluss von ausgewählten Dispositionsparametern auf das Ergebnis von Leistungsuntersuchungen.* In: ETR-Eisenbahntechnische Rundschau, Nr. 7+8, Jr. 64, 2015, pp. 20-23.

[Martin and Li, 2014]

Martin, Ullrich; Li, Xiaojun: *Entwicklung einer simulationsbasierten Methodik zur ursachenbe-zogenen Engpassbewertung komplexer Gleisstrukturen in spurgeführten Verkehrssystemen unter Berücksichtigung stochastischer Bedingungen (EPSUR).* Neues verkehrswissenschaft-liches Journal – NVJ, Verkehrswissenschaftliches Institut Stuttgart, 2014

[Martin, 2014]

Martin, Ullrich: *Performance Evaluation.* In: Ingo Arne Hansen, Jörn Pachl (eds.): Railway Timetabling and Traffic, Eurailpress, Hamburg, 2014

[Martin and Chu, 2012]

Martin, U.; Chu, Z.: *Direkte experimentelle Bestimmung der maximalen Leistungsfähigkeit bei Leistungsuntersuchungen im spurgeführten Verkehr.* DFG Project (MA 2326/6-1), Stuttgart, 2012

[Martin and Schmidt, 2010]

Martin, Ullrich; Schmidt, Christine: *Erhöhung der Effektivität und Transparenz bei Leistungsuntersuchungenmit Simulationsverfahren.* In: ETR, Nr. 7+8, 2010, S. 463–468.

[Martin et al. 2008a]

Martin, Ullrich; Li, Xiaojun; Schmidt, Christine: *PULEIV Projektbericht. Allgemeingültiges Verfahren zur praxisorientierten Bestimmung des Leistungsverhaltens von Eisenbahninfrastrukturen (unveröffentlicht).* Im Auftrag der DB Netz AG., 2008

[Martin et al. 2008c]

Martin, Ullrich; Li, Xiaojun; Schmidt, Christine et al.: *PULEIV Benutzerhandbuch.* Version 1.0.7 (unveröffentlicht), 2008

[Martin et al. 2008d]

Martin, Ullrich; Li, Xiaojun; Schmidt, Christine et al.: *PULEIV Anwenderleitfaden.* Version 1.0.7 (unveröffentlicht), 2008

[Martin, 2002]

Martin, Ullrich: *Deviation Management in Rail Operation.* In: Networks for Mobility, Proceedings of the International Symposium, 2002

[Martin, 1995]

Martin, Ullrich: *Verfahren zur Bewertung von Zug- und Rangierfahrten bei der Disposition.* Dissertation. TU Braunschweig, 1995

[Oetting, 2010]

Oetting, Andreas: *Automatic Timetable Modification for Entire Networks*, Network for Mobility, 5th International Symposium, Stuttgart, Germany, 2010

[OpenTrack, 2016]

OpenTrack: *OpenTrack – Simulation of Railway Networks, 2016.*

[Pachl, 2014]

Pachl, Jörn: *Timetable Design Principles.* In: Ingo Arne Hansen, Jörn Pachl (eds.): Railway Timetabling & Operations, Eurailpress, Hamburg, 2014

[Pachl, 2002]

Pachl, Jörn: *Railway Operation and Control.* VTD Rail Publishing, 2002

[PT1, 2016]

Public Transportation and Railway Operation. Universität Stuttgart, Lecture Note, Winter Semester, 2016

[Radtke, 2014]

Radtke, A.: *Infrastructure Modeling*, In: Ingo Arne Hansen, Jörn Pachl (eds.): Railway Timetabling & Operations, Eurailpress, Hamburg, 2014

[RMCon, 2016]

RMCon: *Railsys Suite – Innovation IT Solution for Railway Transport*, 2016

[RMCon, 2007]

RMCon Software: Railsys-7.6.12, User Manual, English Version, pp. 385-386, 2007

[Samà et al., 2016]

Samà, M.; Pellegrini, P.; D'Ariano, A.; Rodriguez, J.; Pacciarelli, D.: *Ant Colony Optimization for the real-time Train Routing Selection Problem.* In: Transportation Research Part B, Vol. 85, pp. 89-108, 2016

[Shaer et al., 2005]

Schaer, Thorsten; Jacobs, Jürgen; Scholl, Susanne; Kurby, Stephan: *DisKon-Laborversion Eines Flexiblen Modularen und Automatischen Dispositionsassistenzsystems.* In: ETR, Nr. 12, 2005, S.809-821

[Siefer, 2008]

Siefer, Thomas: Simulation. In: Ingo Arne Hansen, Jörn Pachl (eds.): Railway Timetable & Traffic Hamburg: Eurailpress, pp. 155-169, 2008

[Vepsalainen, 1984]

Vepsalainen, Ari P. J: *State Dependent Priority Rules for Scheduling.* Carnegie-Mellon University, Graduate School of Industrial Administration and the Robotics Institute, 1984

[VIA-Con, 2016]

VIA Consulting and Development: *LUKS – Analysis of Lines and Junctions,* (16.12.2016). <http://www.via-con.de/en/development/luks>, Accessed 16.December 2016.